Applied Mathematical Sciences
Volume 105

Applied Mathematical Sciences

(continued following index)

Stephen Wiggins

Normally Hyperbolic Invariant Manifolds in Dynamical Systems

With 22 Illustrations

With the Assistance of
György Haller and Igor Mezić

Springer-Verlag

New York Berlin Heidelberg London Paris
Tokyo Hong Kong Barcelona Budapest

Stephen Wiggins
Applied Mechanics Department 104-44
California Institute of Technology
Pasadena, CA 91125
USA

Editors

F. John
Courant Institute of
  Mathematical Sciences
New York University
New York, NY 10012
USA

J.E. Marsden
Department of
  Mathematics
University of California
Berkeley, CA 94720
USA

L. Sirovich
Division of
  Applied Mathematics
Brown University
Providence, RI 02912
USA

Mathematics Subject Classification (1991): 34K20, 34A30, 39A10

Library of Congress Cataloging-in-Publication Data
Wiggins, Stephen.
    Normally hyperbolic invariant manifolds in dynamical systems/
Stephen Wiggins.
        p.   cm. — (Applied mathematical sciences; v. 105)
    Includes bibliographical references and index.
    ISBN 0-387-94205-X (New York: acid-free). — ISBN 3-540-94205-X
(Berlin: acid-free):
        1. Differentiable dynamical systems.   2. Ordinary differential equations.
    3. Invariant manifolds.   I. Title.   II. Series: Applied mathematical
    sciences (Springer-Verlag Inc.): v. 105.
    QA1.A647   vol. 105
    [QA614.8]
    510 s — dc20
    [514′.74]   94-8078

Printed on acid-free paper.

Production managed by Natalie Johnson; manufacturing supervised by Gail Simon.
Photocomposed using the author's LaTex files.
Printed and bound by Edwards Brothers, Inc., Ann Arbor, MI.
Printed in the United States of America.

9 8 7 6 5 4 3 2 1

ISBN 0-387-94205-X Springer-Verlag New York Berlin Heidelberg
ISBN 3-540-94205-X Springer-Verlag Berlin Heidelberg New York

Preface

Invariant manifolds have a long history in dynamical systems theory—historical descriptions can be found in Fenichel [1971] and Hirsch, Pugh, and Shub [1977]. In the late 1960s and throughout the 1970s, this theory began to assume a very general and well-developed form that is embodied in the works of Fenichel [1971], [1974], [1977] and Hirsch, Pugh, and Shub [1977]. As applications of dynamical systems theory flowered throughout the 1980s "invariant manifold theorems" became standard tools for applied mathematicians, physicists, engineers, and virtually anyone working on nonlinear problems from a geometric viewpoint.

The purpose of this book is to provide detailed proofs of three important invariant manifold theorems; the persistence and smoothness theorem for overflowing invariant manifolds; the unstable manifold theorem for overflowing invariant manifolds; and the foliation theorem for unstable manifolds of overflowing invariant manifolds—as well as indicate how they can be used in applications. I view these three theorems as absolutely fundamental for the analysis of nonlinear dynamical systems from the geometric viewpoint. For this reason, I feel that it will be useful to present their proofs in a unified, detailed, and pedagogic setting, while at the same time pointing out how they may be used in applications. I follow the setup of Fenichel closely—indeed, the persistence and smoothness theorem and the unstable manifold theorem for overflowing invariant manifolds are directly from Fenichel [1971]. In some sense, the foliation theorem for unstable manifolds of overflowing invariant manifolds presented here is new, but the debt to Fenichel [1974] is great.

I would like to thank my assistants, György Haller and Igor Mezić, for helping with the preparation of this book. They provided invaluable help in checking the details of the proofs of the many lemmas, propositions, corollaries, and theorems, as well as the general setup, and in many cases they came up with more improved versions of their own.

The bulk of this book was written at the Fields Institute during January through March of 1993, and I would like to thank the organizers of the Dynamical Systems and Bifurcation Theory year at the Fields Institute—John Chadam, Leon Glass, Bill Langford, Jerry Marsden, and Bill Shadwick—for making it possible for me to spend time at the Fields Institute. I am grateful to my wife Meredith for making my visit to the Fields Institute possible by taking over all of my household and childcare duties. At the same time, I would like to acknowledge the support of my colleague in Applied Mechanics at Caltech, Jim Knowles. Jim took over my teaching load on top of his own, so that I could spend three months at the Fields Institute. I would like to also acknowledge the wonderful help and support of the Fields Institute staff: Sue Embro, Sherri Albers, Ron Hosler, Judy Motts, Jacqua Taylor, and Sandra Valeriote. Their efforts, in too many areas to mention, made my visit both pleasant and productive.

The Fields Institute for Research in Mathematical Sciences is supported by the Ontario Ministry of Education and Training and the Natural Sciences and Engineering Research Council of Canada.

I would also like to thank Jerry Marsden for reading an early draft of the book and for providing editorial assistance with references related to background and applications. Dave McLaughlin and Charles Li made a number of useful and insightful comments on the chapter on foliations of unstable manifolds.

Finally, I would like to acknowledge the support of the Office of Naval Research and the National Science Foundation over the past several years which has made the preparation of this book possible.

Contents

1

Introduction, Motivation, and Background

In this chapter we want to describe and motivate some aspects of the theory of invariant manifolds which we will explore throughout the rest of this book, as well as give a brief survey of the field and the breadth of applications of invariant manifold methods and ideas throughout science and engineering. Roughly speaking, an invariant manifold is a surface contained in the phase space of a dynamical system that has the property that orbits starting on the surface remain on the surface throughout the course of their dynamical evolution in one or both directions of time; i.e., an invariant manifold is a collection of orbits which form a surface. If the surface has a boundary, then trajectories can leave the surface by crossing the boundary. Additionally, under certain conditions, the set of orbits which approach or recede from an invariant manifold M asymptotically in time are also invariant manifolds which are called the stable and unstable manifolds, respectively, of the invariant manifold. Moreover, the stable and unstable manifolds may be *foliated* or *fibered* by lower dimensional submanifolds having the property that trajectories starting on these *fibers* satisfy certain asymptotic growth rate conditions. Many of these features persist under perturbation, a property which is useful for the development of many types of global perturbation methods. We begin by considering areas of applications where ideas from invariant manifold theory have played an important role.

1.1 Examples of the Use of Invariant Manifold Ideas and Methods in Applications

The importance of the notion of "invariant manifolds" in the study of applied dynamical systems has increased dramatically in recent years. Invariant manifolds have proven to be an important tool for thinking about and formulating problems arising in a remarkably broad range of applications. Below, we mention just a few of the applications in a variety of areas in science and engineering. This list is not intended to be exhaustive or complete, but rather illustrative of the extent to which invariant manifolds permeate the literature of applied dynamics.

Fluid Mechanics–Kinematics. In the last 10 years there has been much work in the study of convective transport and mixing problems in fluid mechanics from the point of view of dynamical systems theory. In this framework the stable and unstable manifolds of hyperbolic invariant sets of the flow field form the spatial "template" which organizes the transport and mixing processes. See, e.g., Ottino [1989] and Wiggins [1992].

Fluid Mechanics–Dynamics. There is much interest in whether or not viscous fluid mechanics is "finite dimensional." One of the ways in which this issue is mathematically formulated is to consider the appropriate fluid dynamical partial differential equations as a dynamical system in infinite dimensional function space. One would then like to show that all trajectories are attracted to a finite dimensional invariant submanifold. See, e.g., Temam [1988] and Constantin et al. [1989].

Theoretical Chemistry–Molecular Reaction Dynamics. Invariant noindent manifolds play an important role as structures in classical phase space that control the transfer or sharing of energy between different vibrational or rotational modes of molecules. See, e.g., Davis [1987] and Gillilan and Ezra [1991].

Ship Dynamics. It has been shown that invariant manifolds form the boundary in phase space between safe rolling motions of a ship and motions that will lead to the ship capsizing. See Falzarano, Shaw, and Troesch [1992].

Rigid Body Dynamics. There has been much work in the past 15 years on the phase space geometry of the rigid body. In this work, invariant manifolds play an important role in developing and describing the geometric structure of phase space. See Abraham and Marsden [1988].

In an engineering application of rigid body dynamics Bruhn and Koch [1991] have considered the rocking response of a rigid structure and have shown that a knowledge of invariant manifolds is important for understanding the dynamics of overturning of the structure.

Chaotic Scattering. In certain potential scattering problems there may be a critical energy value at which there is a transition between regular and chaotic scattering trajectories. It has been shown by Bleher, Grebogi, and Ott [1990] that an important feature for understanding this phenomenon is the geometry of invariant manifolds in phase space.

Solid Mechanics—The Formation of Fine Structure. Ball et al. [1991] have studied models of mechanical systems that dissipate energy as time increases. The models derive from potential energy functions that do not obtain a minimum, but rather possess minimizing sequences with finer and finer structure that converge weakly to non-minimizing states. Their models are designed to mimic the dynamical development of the fine structure that is observed in certain material phase transitions. Invariant manifold concepts play an important role in their analysis and description of this limiting behavior.

Condensed Matter Physics–Mesoscale Structures. Wagenhuber et al. [1992] study the classical dynamics of a charged particle in a two-dimensional lattice-periodic potential with a perpendicular magnetic field. In this work they find that the onset of one-dimensional diffusive motion of the charged particle is due to the intersection of the stable and unstable manifolds of different hyperbolic invariant sets.

Mathematical Biology. Terman [1992] has studied the transition from bursting to continuous spiking in excitable membrane models such as those describing electrical activity in pancreatic β-cells. He has shown that invariant manifolds lie at the heart of the mechanism that gives rise to the the transition between these two types of behavior.

Oceanography. Allen, Samelson, and Newberger [1991] study a model for wind-forced quasi-geostrophic flow over a continental margin with variable topography. Their analysis shows that the stable and unstable manifolds of a normally hyperbolic invariant manifold are a mechanism for chaotic behavior in their model. The chaos is characterized by a flow that switches randomly between three oscillatory free wave regimes and could be connected to the observed lack of regularity of coastal ocean currents.

Meteorology. Kopell [1985] has shown that under certain conditions the "quasigeostrophic states" form an attracting invariant manifold for a finite-dimensional system of atmospheric equations.

Bubble Dynamics. Kang and Leal [1990] have studied the dynamics of bubbles under external straining flows and have shown that the process of bubble breakup is controlled by invariant manifolds.

Plasma Physics. Petschel and Geisel [1991] have studied the guiding center motion of a charged particle in the presence of three electrostatic waves. This system models certain features of charged particle transport in toroidal plasma fusion. They found that the mechanism for the onset of large scale stochasticity is related to the geometrical properties and interactions of three different invariant manifolds.

Control Theory. Invariant manifold notions are playing an increasingly important role in the development of nonlinear control theory. Chapter 11 (and the references at the end of this chapter) of the recent book of Nijmeijer and van der Schaft [1990] gives a good picture of this work.

Electronic Circuit Theory. Odyniec and Chua [1983], [1985] have analyzed the Josephson-junction circuit from the point of view of invariant manifold theory. Their analysis enabled them to explain a variety of experimentally observed phenomena such as hysteresis and the existence of voltage steps.

Celestial Mechanics. Recently Touma and Wisdom [1993] have discovered that the obliquity of the planet Mars undergoes large chaotic variations. They show that these variations occur as the system evolves in a chaotic zone associated with a secular spin-orbit resonance. Invariant manifolds play an important role in describing this chaotic zone.

Chaos Theory. The most important and widespread mechanism for the generation of chaotic behavior in deterministic dynamical systems is homoclinic and heteroclinic phenomena, which is characterized by the intersection of the stable and unstable manifolds of normally hyperbolic invariant manifolds. See, e.g., Wiggins [1988] and Palis and Takens [1993]. Strange attractors are often characterized as the closure of the unstable manifold of some hyperbolic invariant set. See Benedicks and Carleson [1991] and Palis and Takens [1993].

Singular Perturbation Theory and Asymptotic Expansions. Foliations of stable and unstable manifolds are becoming an increasingly important tool for studying singular perturbation problems as well as in constructing asymptotic expansions that approximate the solutions of these problems. See, e.g., Fenichel [1979], Kopell [1979] [1985], and Lin [1989].

1.2   Issues and Methods

Next we will survey some background in the mathematical development of the subject of invariant manifold theory. Insight into the history of the work done in this subject can be obtained by focussing on two specific questions. What are the types of results and theorems that have been proven? What methods have been used to obtain those results? We begin with the first question.

We will organize the types of results and theorems that have been proven concerning invariant manifolds by considering them in the context of the four main issues that arise in applications.

1.2.1 EXISTENCE OF INVARIANT MANIFOLDS

Invariant manifold theory begins by *assuming* that some "basic" invariant manifold exists. The theory is then developed from this point onwards by building upon this basic invariant manifold. In particular, one is interested in the construction of stable, unstable, and center manifolds associated with these basic invariant manifolds. The types of basic invariant manifolds typically considered are:

1. Equilibrium points,

2. Periodic orbits,

3. Quasiperiodic or almost periodic orbits (invariant tori).

These manifolds all share an important property. Namely, they all admit a global coordinate description. In this case the dynamical system is typically subjected to a "preparatory" coordinate transformation that serves to localize the dynamical system about the invariant manifold. This amounts to deriving a normal form in the neighborhood of an invariant manifold, and it greatly facilitates the various estimates that are required in the analysis. Sacker [1969], Fenichel [1971], and Hirsch, Pugh, and Shub [1977] were among the first to consider general invariant manifolds that are not described as graphs and required an atlas of local coordinate charts for their description. This is an important generalization since in recent years invariant manifolds that cannot be expressed globally as a graph have arisen in applications. See Wiggins [1990], Hoveijn [1992], and Haller and Wiggins [1992] for applications where invariant spheres arise.

1.2.2 THE PERSISTENCE AND DIFFERENTIABILITY OF INVARIANT MANIFOLDS UNDER PERTURBATION

The question of whether or not an invariant manifold persists under perturbation and, if so, if it maintains, loses, or gains differentiabilty is also important. In considering these issues it is important to characterize the stability of the *unperturbed* invariant manifold. This is where the notion of *normal hyperbolicity* arises. Roughly speaking, a manifold is normally hyperbolic if, under the dynamics *linearized about the invariant manifold*, the growth rate of vectors transverse to the manifold dominates the growth rate of vectors tangent to the manifold. For equilibrium points, these growth rates can be characterized in terms of eigenvalues associated with the linearization at the equilibria that are not on the imaginary axis, for periodic orbits, these growth rates can be characterized in terms of the Floquet multipliers associated with the linearization about the periodic orbit that are not on the unit circle. For invariant tori or more general invariant manifolds these growth rates can be characterized in terms of exponential dichotomies (see Coppel [1978] or Sacker and Sell [1974]) or by the notion of generalized Lyapunov-type numbers (Fenichel [1971]), which is the approach that we will take in this book. A question of obvious importance for applications is how does one *compute* whether or not an invariant manifold is normally hyperbolic? The answer is not satisfactory. For equilibria, the problem involves finding the eigenvalues of a square matrix—an algebraic problem. For invariant manifolds on which the dynamics is nontrivial, the issues are more complicated, and they are dealt with in this book. However, one important class of dynamical systems which may have nontrivial invariant manifolds on which the dynamics is also nontrivial, are integrable Hamiltonian systems. See Wiggins [1988] for examples. Finally, we want to alert the reader to an important characteristic of normal hyperbolicity that is of some importance for understanding the scope of possible applications. Namely, it is insensitive to the form of the dynamics on the invariant manifold, provided the dynamics transverse to the invariant manifold is dominant in the sense of normal hyperbolicity. Heuristically, one could think of the dynamics on the invariant manifold as being "slow" as compared to the "fast" dynamics off the invariant manifold. Hence, the dynamics on the invariant manifold could even be chaotic.

Characterizing growth rates in the fashion described above requires knowledge of the linearized dynamics near orbits on the invariant manifold as $t \to +\infty$ or $t \to -\infty$. Hence, if the invariant manifold has a boundary (which an equilibrium point, periodic orbit, or invariant torus *does not* have), then one must understand the nature of the dynamics at the boundary. Notions such as *overflowing invariance* or *inflowing invariance* are developed to handle this. Invariant manifolds with boundary arise very often in applications, see Wiggins [1988].

1.2.3 BEHAVIOR NEAR AN INVARIANT MANIFOLD— STABLE, UNSTABLE, AND CENTER MANIFOLDS

A "stable manifold theorem" asserts that the set of points that approach an invariant manifold at an exponential rate as $t \to +\infty$ is an invariant manifold in its own right. The exponential rate of approach is inherited from the linearized dynamics as the stable manifold is constructed as a graph over the linearized stable *subspace* or *subbundle*. An "unstable manifold theorem" asserts similar behavior in the limit as $t \to -\infty$. Obviously, one may have problems with both of these concepts if the invariant manifold has a boundary.

The notion of a center manifold is more subtle. For equilibrium points and periodic orbits, a center manifold is an invariant manifold that is tangent to the linearized subspace corresponding to eigenvalues on the imaginary axis, and floquet multipliers on the unit circle, respectively. In contrast to the situation with stable and unstable manifolds, the asymptotic behavior of orbits in the nonlinear center manifold may be very different than the asymptotic behavior of orbits in the linearized center subspaces, under the linearized dynamics.

Questions related to persistence and differentiability of stable, unstable, and center manifolds also arise.

1.2.4   More Refined Behavior Near Invariant Manifolds—Foliations of Stable, Unstable, and Center Manifolds

One may be interested in which orbits in the stable manifold approach the invariant manifold at a specified *rate*. Under certain conditions these orbits may lie on submanifolds of the stable manifold which are not invariant, but make up an invariant family of submanifolds that *foliate* the stable manifold. A similar situation may hold for the unstable manifold. Moreover, this foliation has the property that points in a *fiber* of the foliation asymptotically approach the trajectory in the invariant manifold that passes through the point of intersection of the fiber with the invariant manifold (the *basepoint* of the fiber). This is a generalization of the notion of *asymptotic phase*, that is familiar from studies of stability of periodic orbits, to arbitrary invariant manifolds. In recent years these foliations have seen many uses in applications. Fenichel [1992] has used them in his development of geometric singular perturbation theory. Kovačič and Wiggins [1992], Haller and Wiggins [1993a], [1993b], and McLaughlin et al. [1993] have used them in the development of new global perturbation methods. Finally, the recent monograph of Kirchgraber and Palmer [1990] proves a number of foliation results and shows how these can be used as coordinates in which the dynamical system becomes linear.

Next we consider the methods by which invariant manifold results have been proven.

1.2.5   The Liapunov–Perron Method

Perron [1928], [1929], [1930] and Liapunov [1947] developed a method for proving the existence of stable and unstable manifolds of a hyperbolic equilibrium point. This method is functional-analytic in nature. In the context of ordinary differential equations, it deals with the integral equation formulation of the ordinary differential equations and constructs the invariant manifolds as a fixed point of an operator that is derived from this integral equation on a function space whose elements have the appropriate interpretation as stable and unstable manifolds. The Liapunov–Perron method has been used in many different situations. The book of Hale [1980] is a good reference point for surveying these applications. In this book Hale surveys the fundamental earlier work of Krylov and Bogoliubov, Bogoliubov and

Mitropolski, Diliberto, Kyner, Kurzweil, and Pliss. In Chapter 7 of Hale [1980] several theorems related to various aspects of invariant manifolds are given which we refer to collectively as the "Hale invariant manifold theorem." These results can be viewed as a generalization and extension of much of the earlier work. Yi [1993a], [1993b] has recently generalized the Hale invariant manifold theorem and has also proven some results on foliations.

Sell [1978] uses the Liapunov–Perron method to prove the existence of stable, unstable, and center manifolds of almost periodic orbits. In the infinite dimensional setting, the method is used by Henry [1981] to prove the existence of stable, unstable and, center manifolds for semilinear parabolic equations. Ball [1973] also has a number of results along these lines. The inertial manifold theory is also developed using this approach (see, e.g., Temam [1988] and Constantin et al. [1989]).

1.2.6   HADAMARD'S METHOD—THE GRAPH TRANSFORM

Hadamard [1901] developed this method to prove the existence of stable and unstable manifolds of a fixed point of a diffeomorphism. He also obtained one of the first foliation results using this method. The graph transform method is more geometrical in nature than the Liapunov–Perron method. In the context of a hyperbolic fixed point, the stable and unstable manifolds are constructed as graphs over the linearized stable and unstable subspaces, respectively—hence the name. Fenichel [1971], [1974], [1977] and Hirsch, Pugh, and Shub [1977] used this method in obtaining their general results on normally hyperbolic invariant manifolds. The infinite dimensional invariant manifold work of Bates and Jones [1989] is also in the spirit of the Hadamard graph transform.

1.2.7   THE LIE TRANSFORM, OR DEFORMATION METHOD

The Lie transform, or deformation method, is a familiar method in dynamics. The idea is to use coordinate transformations that are the time-one map of Hamiltonian flows that are constructed in such a way as to accomplish the desired goal. This method is used in canonical perturbation theory and, along these lines, it has recently played a role in proofs of the KAM theorem (Benettin et al. [1984]) and Nekhoroshev's theorem (Lochak [1992], Pöschel [1993]). In a wider context, it has been used in new proofs of the Frobenius theorem and in Darboux's theorem (Abraham and Marsden [1978]).

The method was first developed in the context of invariant manifold theory by Marsden and Scheurle [1987] and used to prove the existence of center-stable and center-unstable manifolds of a fixed point of a diffeomorphism.

1.2.8 IRWIN'S METHOD

This is a method developed by Irwin [1970] that is based on an application of the implicit function theorem in the function space of sequences (that can be thought of as orbits of a discrete dynamical system) that have the appropriate asymptotic behavior. Irwin used this method to prove the existence of the stable and unstable manifolds of a fixed point of a discrete time-dynamical system. A streamlined proof of this result was later given by Wells [1976]. Subsequently, the method was developed by Irwin [1980] and Chow and Hale [1982] in the situation where the fixed point had a center manifold. New streamlined proofs of these results, also valid for center manifolds, have recently been given by de la Llave and Wayne [1993].

1.3 Motivational Examples

All of these aspects of invariant manifolds will be developed in great detail in this book; however, first we begin with a familiar, motivational, example.

Example 1.3.1.

We consider a nonlinear, autonomous ordinary differential equation defined on $\mathbb{R}^n$,

$$\dot{x} = f(x), \qquad x(0) = x_0, \qquad x \in \mathbb{R}^n, \tag{1.1}$$

where $f : \mathbb{R}^n \to \mathbb{R}^n$ is at least C^1. We make the following assumptions on (1.1):

1. Equation (1.1) has a fixed point at $x = 0$, i.e., $f(0) = 0$.

2. $Df(0)$ has $n - k$ eigenvalues having positive real parts and k eigenvalues having negative real parts.

Thus, (1.1) possesses a particularly trivial type of invariant manifold, namely, the fixed point at $x = 0$. Let us now study the nature of the linear system obtained by linearizing (1.1) about the fixed point $x = 0$. We denote the linearized system by

$$\dot{\xi} = Df(0)\,\xi, \qquad \xi \in \mathbb{R}^n, \tag{1.2}$$

and note that the linearized system possesses a fixed point at the origin. Let $v^1, \ldots, v^{n-k}$ denote the generalized eigenvectors corresponding to the eigenvalues having positive real parts, and $v^{n-k+1}, \ldots, v^n$ denote the generalized eigenvectors corresponding to the eigenvalues having negative real parts. Then the linear subspaces of $\mathbb{R}^n$ defined as

$$
\begin{aligned}
E^u &= \operatorname{span}\{v^1, \ldots, v^{n-k}\}, \\
E^s &= \operatorname{span}\{v^{n-k+1}, \ldots, v^n\},
\end{aligned}
\tag{1.3}
$$

are invariant manifolds for the linear system (1.2) which are known as the unstable and stable subspaces, respectively. E^u is the set of points such that orbits of (1.2) through these points approach the origin asymptotically in negative time, and E^s represents the set of points such that orbits of (1.2) through these points approach the origin asymptotically in positive time. (Note: These statements are not hard to prove, and we refer the reader to Arnold [1973] or Hirsch and Smale [1974] for a thorough discussion of linear, constant coefficient systems.) We represent this situation geometrically in Fig. 1.1.

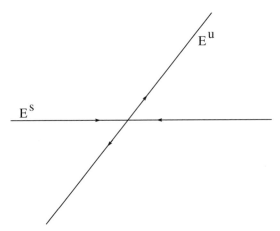

FIGURE 1.1. Stable and Unstable Subspaces of the Linearized System.

The question we now ask is, What is the behavior of the nonlinear system (1.1) near the fixed point $x = 0$? We might expect that the linearized system should give us some indication of the nature of the orbit structure near the fixed point of the nonlinear system since the fact that none of the eigenvalues of $Df(0)$ have a zero real part implies that near $x = 0$ the flow of (1.1) is dominated by the flow of (1.2). (Note: Fixed points of vector fields which have the property that the eigenvalues of the matrix associated with the linearization of the vector field about the fixed point have nonzero real parts are called *hyperbolic* fixed points.) Indeed, the stable and unstable manifold theorem for hyperbolic fixed points tells us that in a neighborhood N of the fixed point $x = 0$ for (1.1), there exists a differentiable (as differentiable as the vector field (1.1)) $n - k$-dimensional surface, $W^u_{loc}(0)$, tangent to E^u at $x = 0$ and a differentiable k-dimensional surface, $W^s_{loc}(0)$, tangent to E^s at $x = 0$ with the properties that orbits of points on $W^u_{loc}(0)$ approach $x = 0$ asymptotically in negative time (i.e., as $t \to -\infty$) and orbits of points on $W^s_{loc}(0)$ approach $x = 0$ asymptotically in positive time (i.e., as $t \to +\infty$). $W^u_{loc}(0)$ and $W^s_{loc}(0)$ are known as the local unstable and stable manifolds, respectively, of $x = 0$. We represent this situation geometrically in Fig. 1.2.

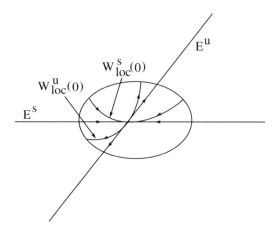

FIGURE 1.2. Phase Space of Eq. (1.1) near $x = 0$.

Let us denote the flow generated by (1.1) as $\phi_t(\cdot)$; then we can define global stable and unstable manifolds of $x = 0$ by using points on the local manifolds as initial conditions:

$$
\begin{aligned}
W^u(0) &= \bigcup_{t \geq 0} \phi_t \left(W^u_{\text{loc}}(0) \right), \\
W^s(0) &= \bigcup_{t \leq 0} \phi_t \left(W^s_{\text{loc}}(0) \right).
\end{aligned} \tag{1.4}
$$

$W^u(0)$ and $W^s(0)$ are called the unstable and stable manifolds, respectively, of $x = 0$. We represent the situation geometrically in Fig. 1.3.

Now suppose we add a small autonomous perturbation, $\epsilon g(x)$, to (1.1) where $g(x)$ is as differentiable as $f(x)$ and $\epsilon \in I \subset \mathbb{R}$ where $I = \{ \epsilon \in \mathbb{R} \mid -\epsilon_0 < 0 < \epsilon_0 \}$. We denote the perturbed system by

$$
\dot{x} = f(x) + \epsilon g(x), \qquad x(0) = x_0, \qquad x \in \mathbb{R}^n. \tag{1.5}
$$

The question we now ask is, How much of the structure of (1.1) is preserved in the perturbed system (1.5)? Specifically, we will be concerned with what happens to the fixed point at the origin and its stable and unstable manifolds.

The fate of the fixed point is easy to determine by a simple application of the implicit function theorem. (Note: Recall that a fixed point of (1.5) is a solution of $f(x) + \epsilon g(x) = 0$.) We will set up the problem for application of the implicit function theorem. Let us consider the function

$$
\begin{aligned}
G : \mathbb{R}^n \times I &\rightarrow \mathbb{R}^n, \\
(x, \epsilon) &\mapsto f(x) + \epsilon g(x).
\end{aligned} \tag{1.6}
$$

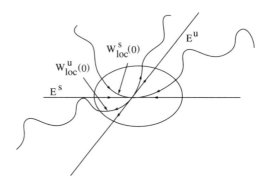

FIGURE 1.3. Global Stable and Unstable Manifolds of $x = 0$.

It is clear that $G(0,0) = 0$, and we wish to determine if there exists a solution of $G(x, \epsilon) = 0$ for (x, ϵ) close to $(0, 0)$. Now the derivative of G with respect to x evaluated at $(x, \epsilon) = (0, 0)$ is given by

$$D_x G(0,0) = D_x f(0). \tag{1.7}$$

By our assumption on the eigenvalues of $Df(0)$ (specifically, there are no zero eigenvalues) it is clear that $\det[D_x G(0,0)] = \det D_x f(0) \neq 0$; thus, by the implicit function theorem there exists a function of ϵ, $\bar{x}(\epsilon)$ (with $\bar{x}(\epsilon)$ as differentiable as $G(x, \epsilon)$), such that

$$G(\bar{x}(\epsilon), \epsilon) = 0 \tag{1.8}$$

for ϵ sufficiently small contained in I. Thus, the fixed point is preserved in the perturbed system, although it may move slightly.

The fate of the unstable and stable manifolds of $x = 0$ follows from the persistence theory for stable and unstable manifolds which we will describe in later chapters. However, for now we will state the consequence of this theory, which tells us that in some neighborhood $\tilde{N}$ containing $x = 0$ and $x = \bar{x}(\epsilon)$ there exist differentiable manifolds $\tilde{W}^u_{\text{loc}}(\bar{x}(\epsilon))$ and $\tilde{W}^s_{\text{loc}}(\bar{x}(\epsilon))$ passing through $\bar{x}(\epsilon)$ with the properties that orbits of points in $\tilde{W}^u_{\text{loc}}(\bar{x}(\epsilon))$ under the perturbed flow approach $x = \bar{x}(\epsilon)$ asymptotically in negative time and orbits of points in $\tilde{W}^s_{\text{loc}}(\bar{x}(\epsilon))$ under the perturbed flow approach $x = \bar{x}(\epsilon)$ asymptotically in positive time. $\tilde{W}^u_{\text{loc}}(\bar{x}(\epsilon))$ and $\tilde{W}^s_{\text{loc}}(\bar{x}(\epsilon))$ have the same dimensions and differentiability as $W^u_{\text{loc}}(0)$ and $W^s_{\text{loc}}(0)$, respectively. Utilizing the flow generated by the perturbed system (1.5) and $\tilde{W}^u_{\text{loc}}(\bar{x}(\epsilon))$ and $\tilde{W}^s_{\text{loc}}(\bar{x}(\epsilon))$ as initial conditions, we can define global unstable and stable manifolds of $x = \bar{x}(\epsilon)$ in exactly the same manner as we defined them for the unperturbed system. See Fig. 1.4 for a geometrical interpretation.

The central theme of this book is to generalize this example to the case where the invariant manifold under consideration is a normally hyperbolic *invariant manifold*.

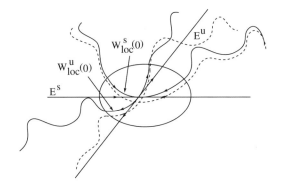

FIGURE 1.4. Perturbed and Unperturbed Structure.

End of Example 1.3.1

We use this simple example to illustrate several points which will arise in the course of this generalization.

1. For the unperturbed equation it is first necessary to locate the invariant manifold. In our simple example the invariant manifold is a fixed point which can be found by solving for the zeros of a system of coupled nonlinear algebraic relations. Locating more general types of invariant manifolds may involve having quite a detailed knowledge of the orbit structure of a nonlinear ordinary differential equation, which, in general, is a formidable task.

2. Once the invariant manifold of the unperturbed system is obtained, it is then necessary to study the linear system obtained by linearizing the unperturbed system about the invariant manifold. This procedure, where the invariant manifold is a fixed point, periodic orbit, or quasiperiodic orbit, is quite familiar; if the unperturbed system is of the form

$$\dot{x} = f(x), \qquad x \in \mathbb{R}^n, \quad f \in C^1, \tag{1.9}$$

with an invariant manifold $\phi(t)$ being a fixed point, periodic orbit, or quasiperiodic orbit, then letting $x(t) = \phi(t) + \xi(t)$, we obtain

$$\dot{x} = \dot{\phi} + \dot{\xi} = f(\phi + \xi) = f(\phi) + Df(\phi)\,\xi + O\left(|\xi|^2\right)$$

or

$$\dot{\xi} = Df(\phi)\,\xi + O\left(|\xi|^2\right) \tag{1.10}$$

since $\dot{\phi} = f(\phi)$ (i.e., ϕ is a solution of (1.9)). If we retain only terms linear in ξ, we obtain the associated linearized system

$$\dot{\xi} = Df(\phi)\,\xi. \tag{1.11}$$

Now if the invariant manifold is more general, such as a surface containing many different orbits of (1.9), then linearizing about the invariant manifold is not a straightforward procedure, especially if the invariant manifold is not globally representable as a graph of a function. In this case, one obtains a collection of linear equations representing the linearized vector field in different "coordinate charts" on the invariant manifold. The techniques for describing the vector field near a general invariant manifold are obtained from the theory of differentiable manifolds which we will describe shortly.

3. Once the linearized system is obtained, it is then necessary to study its stability. This information will allow us to determine the dimension of the stable and unstable manifolds of the invariant manifold as well as the persistence and smoothness properties of the structure under perturbation. In general, this is a formidable task since the coefficients of the linear system may have a complicated time dependence. There are two approaches to the problem: One involves the computation of generalized Lyapunov-type numbers (this is the approach we shall take) and the other a consideration of exponential dichotomies (see Coppel [1978] and Sacker and Sell [1974]).

Next we want to motivate and introduce the notion of a *foliation* or *fibration* of the stable and unstable manifolds of an invariant manifold. We begin with an example from McLaughlin et al. [1993].

Example 1.3.2.

Consider the following planar vector field:

$$\dot{x} = -\epsilon (x+y), \quad 0 \leq \epsilon << 1,$$
$$\dot{y} = -y.$$

The trajectory passing through the point $p_1 \equiv (x_1, y_1)$ at $t = 0$ is given by

$$x^{(\epsilon)}(t; p_1) = \left(x_1 - \frac{\epsilon}{1-\epsilon} y_1\right) \exp(-\epsilon t) + \left(\frac{\epsilon}{1-\epsilon} y_1\right) \exp(-t),$$
$$y^{(\epsilon)}(t; p_1) = y_1 \exp(-t). \qquad (1.12)$$

One can readily verify that the set

$$M = \{(x, y) \mid y = 0\},$$

is an invariant manifold and the rest of the x-y plane is the stable manifold of M, $W^s(M)$. The dynamics on M is given by

$$\dot{x} = -\epsilon x.$$

In Fig. 1.5 we show the trajectories for $\epsilon > 0$ and $\epsilon = 0$. Clearly, the trajectories in the two cases do *not* remain close for all time. In particular, near M they diverge (except, of course, on the y-axis).

One can see from inspection of (1.12) that there are two growth rates associated with trajectories. The x components of two trajectories approach each other at the rate $\exp(-\epsilon t)$ and the y components approach each other at the rate $\exp(-t)$. We now ask the following question:

What is the nature of the set of points in $W^s(M)$ such that trajectories through these points approach the trajectory through p_1 at the fastest rate as $t \to \infty$?

The answer to this question can be obtained through a computation. Consider the trajectory through any other point, say p_2, at $t = 0$. The difference in the two trajectories through p_1 and p_2 is given by

$$x(t; p_2) - x(t; p_1) = \left[\left(x_2 - \frac{\epsilon}{1-\epsilon} y_2\right) - \left(x_1 - \frac{\epsilon}{1-\epsilon} y_1\right)\right] \exp(-\epsilon t)$$
$$+ \frac{\epsilon}{1-\epsilon} (y_2 - y_1) \exp(-t),$$
$$y(t; p_2) - y(t; p_1) = (y_2 - y_1) \exp(-t).$$

By inspection, these two trajectories will approach each other at the fastest rate as $t \to +\infty$ provided the coefficient in front of the $\exp(-\epsilon t)$ term is required to vanish. This gives the equation

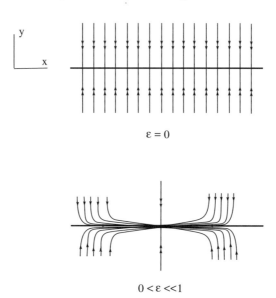

FIGURE 1.5. Trajectories for $\epsilon = 0$ and $0 < \epsilon << 1$.

$$\left\{ x_2 = x_1 + \frac{\epsilon}{1 - \epsilon}(y_2 - y_1), \ y_2 \in \mathbb{R} \right\},$$

which we refer to as the *stable fiber through* p_1. So we see that the answer to our question is that the set of points that approach p_1 at the fastest rate is a smooth curve that is also differentiable in ϵ (for $\epsilon \neq 1$). This fiber is depicted in Fig. 1.6.

The consideration of particular growth rates are often motivated by particular geometrical features of the given vector field. In this example we see that the growth rate $\exp(-\epsilon t)$ is related to the "slow" dynamics on M and the growth rate $\exp(t)$ is related to the fast dynamics transverse to M. In general, the existence of an invariant manifold makes it natural to consider growth rates tangential to the manifold versus growth rates transverse to the manifold. Moreover, the particularly simple, global coordinate representation of the invariant manifold in this example made this growth rate comparison quite simple. This situation will be generalized later to the case of an arbitrary invariant manifold M. We will want to characterize the dynamics on M as "slow" compared to the dynamics off of M. The notion of *generalized Lyapunov-type numbers* will be used for this purpose.

An important use of the fibers is to "relate or connect" the fast dynamics off of M with the slow dynamics on M, over semi-infinite time scales. For this purpose it will be most useful to label each fiber by the point at which

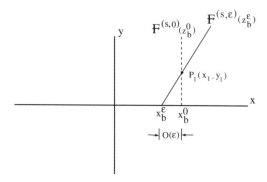

FIGURE 1.6. Stable Fiber.

it intersects M, in this example the x-axis. This point is referred to as the "base point" of the fiber and is obtained by choosing p_1 to be on M, which in this case gives

$$z_b^\epsilon \equiv \left(x_b^\epsilon = x_2 - \frac{\epsilon}{1-\epsilon} \, y_2; \; y_b^\epsilon = 0 \right).$$

Thus, the equation for the fiber passing through this base point is given by

$$\mathcal{F}^{s,\epsilon}(z_b^\epsilon) = \left\{ \left(x = x_b^\epsilon + \frac{\epsilon}{1-\epsilon} \, y; \; y \right), \; y \in \mathbb{R} \right\},$$

where we have dropped the subscript "2" since the point p_2 is arbitrary.

We can also ask a slightly different question:

What is the nature of the set of points in $W^s(M)$ such that trajectories through these points approach the trajectory through $p \in M$ at the fastest rate as $t \to \infty$?

We readily see that the answer to this question is the stable fiber with basepoint equal to the point $p \in M$.

It is important to realize that the one-dimensional fiber is not an invariant manifold (or else it would be a solution trajectory). Rather, it has the property that all points on the fiber, viewed as initial conditions of trajectories, approach the trajectory on M through the basepoint of the fiber. This makes precise the statement that the fibers "relate or connect" the fast dynamics with the slow dynamics from above. More heuristically, the fibers can be thought of as curves of initial conditions that "shadow" specific trajectories on M. Also notice that, in this example, the fiber is differentiable in ϵ, as well as differentiable with respect to the basepoint.

Moreover, we can represent $W^s(M)$ as a union of fibers as follows:

$$W^s(M) = \bigcup_{z_b^\epsilon \in M} \mathcal{F}^{s,\epsilon}(z_b^\epsilon).$$

We say that the two-dimensional $W^s(M)$ is *foliated* by one-dimensional curves (actually, lines), and the curves correspond to initial conditions that approach each other at the fastest rate. An important use of fibers in this setting, as opposed to trajectories, is for developing perturbation techniques that are valid over semi-infinite time intervals in systems with different time scales.

————————————————————————————— End of Example 1.3.2

We next give an example that shows that the stable manifold of an invariant manifold need not be foliated in this manner.

Example 1.3.3.
    Consider the following planar vector field:

$$\dot{x} = -x,$$
$$\dot{y} = -y.$$

The invariant manifold situation is the same as that in the previous example; namely,

$$M = \{(x,y) \,|\, y = 0\}$$

is an invariant manifold and the rest of the x-y plane is the stable manifold of M, $W^s(M)$. The trajectory through a given point $p_1 = (x_1, y_1)$ at $t = 0$ is given by

$$x(t) = x_1 e^{-t},$$
$$y(t) = y_1 e^{-t}.$$

Now we ask the same question as in Example 1.3.2.

What is the nature of the set of points in $W^s(M)$ such that trajectories through these points approach the trajectory through p_1 at the fastest rate as $t \to \infty$?

Choose any other point $p_2 = (x_2, y_2)$, and a simple calculation shows that

$$x(t; p_1) - x(t; p_2) = = (x_1 - x_2)\, e^{-t},$$
$$y(t; p_1) - y(t; p_2) = = (y_1 - y_2)\, e^{-t}.$$

From this calculation we see that the *every* point in $W^s(M)$ approaches p_1 at the same rate. Thus, $W^s(M)$ is *not* foliated by one-dimensional submanifolds with the same properties as in the previous example.

_____
                                    End of Example 1.3.3

The difference in Example 2 and Example 3 is that in Example 2 the rate of expansion and contraction of trajectories on M is smaller than the rate of expansion and contraction of trajectories off M if we take ϵ small. In Example 3, these rates are the same. The notion of *generalized Lyapunov-type numbers* introduced in the next chapter will allow us to quantify this idea.

2

Background from the Theory of Differentiable Manifolds

Before discussing the general theory of invariant manifolds, we need to give some background material from differential geometry. More specifically, we will need to understand the definition of a differentiable manifold, the tangent space at a point, the tangent bundle, and the derivatives of maps defined on differentiable manifolds. These notions will be crucial for discussing the dynamics near an invariant manifold. We will not develop these concepts in the most abstract or mathematically crisp manner, but rather along the lines where they occur frequently in applications. In applications involving the modeling of the dynamics of some physical system, we typically choose certain quantities describing various aspects of the system and write down equations describing the time evolution of these quantities. These quantities constitute the phase space of the system with invariant manifolds often arising as surfaces in the phase space. Consequently, we choose to develop the concept of a differentiable manifold as a surface embedded in $\mathbb{R}^n$ (loosely following the exposition of Milnor [1965] and Guillemin and Pollack [1974]) and refer the reader to any differential geometry textbook for the abstract development of the theory of differentiable manifolds (e.g., a standard and very thorough textbook is Spivak [1979]). Our approach will allow us to bypass certain set-theoretic and topological technicalities since our manifolds will inherit much structure from $\mathbb{R}^n$, whose topology is relatively familiar. Additionally, it is hoped that this approach will appeal to the intuition of the reader who has little or no experience with the subject of differential geometry.

2.1 The Definition of a "Manifold" and Examples

We begin by defining the derivative of a map defined on an arbitrary subset of $\mathbb{R}^n$.

Definition 2.1.1 *Consider a C^r map $f: X \to \mathbb{R}^m$ where X is an arbitrary subset of $\mathbb{R}^n$. f is said to be C^r on X if for every point $x \in X$ there exists an open set $U \subset \mathbb{R}^n$ containing x and a C^r map $F: U \to \mathbb{R}^m$ such that $f = F$ on $U \cap X$.*

Definition 2.1.2 *A map* $f: X \to Y$ *of subsets of two Euclidean spaces is called a C^r diffeomorphism if it is one-to-one and onto and if the inverse map* $f^{-1}: Y \to X$ *is also C^r.*

Initially, this definition may seem a bit strange; however, now we give some examples that show that it is quite natural and typically the way one thinks in such situations.

Example 2.1.1.
Consider $S^1 \subset \mathbb{R}^2$ and the map

$$
\begin{aligned}
f : S^1 &\to \mathbb{R}^1, \\
(x, y) &\to x.
\end{aligned}
\tag{2.1}
$$

A smooth extension of this map to all of $\mathbb{R}^2$ is given by the projection map

$$
\begin{aligned}
F : \mathbb{R}^2 &\to \mathbb{R}^1, \\
(x, y) &\to x.
\end{aligned}
\tag{2.2}
$$

————————————————————————— End of Example 2.1.1

Example 2.1.2.
Consider $S^1 \subset \mathbb{R}^2$ and the map

$$
\begin{aligned}
f : S^1 &\to \mathbb{R}^1, \\
(x, y) &\to \frac{1}{x^2 + y^2}.
\end{aligned}
\tag{2.3}
$$

This map extends to the smooth map

$$
F : \mathbb{R}^2 - \{0\} \to \mathbb{R}^1.
\tag{2.4}
$$

————————————————————————— End of Example 2.1.2

We are now in a position to give the definition of a differentiable manifold.

Definition 2.1.3 *A subset $M \subset \mathbb{R}^n$ is called a C^r manifold of dimension m if it possesses the following two structural characteristics:*

1. *There exists a countable collection of open sets* $V^\alpha \subset \mathbb{R}^n$, $\alpha \in$
   A where A is some countable index set, with $U^\alpha \equiv V^\alpha \cap M$ *such*
   that $M = \bigcup_{\alpha \in A} U^\alpha$.

2. *There exists a* C^r *diffeomorphism* x^α *defined on each* U^α *which maps*
   U^α *onto some open set in* $\mathbb{R}^m$.

We make the following remarks regarding Definition 2.1.3:

1. The sets U^α are often called relatively open sets, i.e., open with respect to M.

2. From (2) of Definition 2.1.3 we see that the degree of differentiability of a manifold is the same as the degree of differentiability of the x^α. This implies a certain compatibility condition which must be satisfied on overlapping charts. More specifically, let $(U^\alpha; x^\alpha)$ and $(U^\beta; x^\beta)$, $\alpha, \beta \in A$, be two charts such that $U^\alpha \cap U^\beta \neq \emptyset$; see Fig. 2.1.

3. A standard terminology is that the pair $(U^\alpha; x^\alpha)$ is called a *chart* for M and the union of all compatible charts, i.e., $\bigcup_{\alpha \in A}(U^\alpha; x^\alpha)$, is called an *atlas* for M.

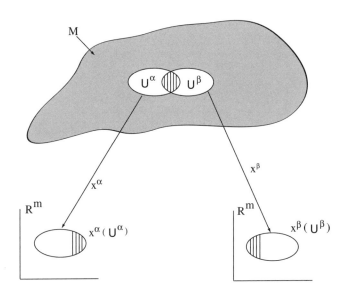

FIGURE 2.1. Coordinate Charts on a Manifold.

Then the region $U^\alpha \cap U^\beta$ can be described by two different coordinatizations, namely, $(U^\alpha \cap U^\beta; x^\alpha)$ and $(U^\alpha \cap U^\beta; x^\beta)$. We denote the coordinate maps by

$$x^\alpha : p \in U^\alpha \cap U^\beta \longrightarrow (x_1^\alpha, \ldots, x_m^\alpha) \in \mathbb{R}^m,$$

$$(2.5)$$

$$x^\beta : p \in U^\alpha \cap U^\beta \longrightarrow (x_1^\beta, \ldots, x_m^\beta) \in \mathbb{R}^m,$$

where $(x_1^\alpha, \ldots, x_m^\alpha)$ and $(x_1^\beta, \ldots, x_m^\beta)$ represent points in the Euclidean space $\mathbb{R}^m$. Now the maps

$$x^\beta \circ (x^\alpha)^{-1} : x^\alpha(U^\alpha \cap U^\beta) \to x^\beta(U^\alpha \cap U^\beta),$$
$$(x_1^\alpha, \ldots, x_m^\alpha) \mapsto \left(x_1^\beta(x_1^\alpha, \ldots, x_m^\alpha), \ldots, x_m^\beta(x_1^\alpha, \ldots, x_m^\alpha)\right),$$

$$x^\alpha \circ (x^\beta)^{-1} : x^\beta(U^\alpha \cap U^\beta) \to x^\alpha(U^\alpha \cap U^\beta),$$
$$(x_1^\beta, \ldots, x_m^\beta) \mapsto \left(x_1^\alpha(x_1^\beta, \ldots, x_m^\beta), \ldots, x_m^\alpha(x_1^\beta, \ldots, x_m^\beta)\right)$$

$$(2.6)$$

represent the change of coordinates from x^α to x^β coordinates and from x^β to x^α coordinates, respectively; the fact that x^α and x^β are C^r diffeomorphisms implies that the maps describing the change of co-ordinates must likewise be C^r diffeomorphisms. [Note: In the change of coordinate maps in formula (2.6) we should more correctly write $\left(x_1^\beta(p(x_1^\alpha, \ldots, x_m^\alpha)), \ldots, x_m^\beta(p(x_1^\alpha, \ldots, x_m^\alpha))\right)$ for the image of $(x_1^\alpha, \ldots, x_m^\alpha)$ under $x^\beta \circ (x^\alpha)^{-1}$ (and similarly for the map $x^\alpha \circ (x^\beta)^{-1}$); however, it is standard and somewhat intuitive to identify points in the manifold with their images in a coordinate chart, especially when the manifold is a sur-face in $\mathbb{R}^n$.] In particular, for $r \geq 1$ we get the familiar requirement on changes of coordinates that the Jacobian matrices

$$\begin{pmatrix} \dfrac{\partial x_1^\beta}{\partial x_1^\alpha} & \cdots & \dfrac{\partial x_1^\beta}{\partial x_m^\alpha} \\ \vdots & & \vdots \\ \dfrac{\partial x_m^\beta}{\partial x_1^\alpha} & \cdots & \dfrac{\partial x_m^\beta}{\partial x_m^\alpha} \end{pmatrix} \quad \text{and} \quad \begin{pmatrix} \dfrac{\partial x_1^\alpha}{\partial x_1^\beta} & \cdots & \dfrac{\partial x_1^\alpha}{\partial x_m^\beta} \\ \vdots & & \vdots \\ \dfrac{\partial x_m^\alpha}{\partial x_1^\beta} & \cdots & \dfrac{\partial x_m^\alpha}{\partial x_m^\beta} \end{pmatrix}$$

be nonsingular on $x^\alpha(U^\alpha \cap U^\beta)$ and $x^\beta(U^\alpha \cap U^\beta)$, respectively.

Heuristically, we see that a differentiable manifold is a set which locally has the structure of ordinary Euclidean space. We now give several exam-ples of manifolds.

Example 2.1.3.

The Euclidean space $\mathbb{R}^n$ is a trivial example of a C^∞ manifold. We take as the single coordinate chart $(i; \mathbb{R}^n)$ where i is the inclusion map; it should be clear that i is infinitely differentiable and, hence, $\mathbb{R}^n$ is a C^∞ manifold.

End of Example 2.1.3

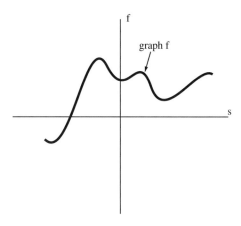

FIGURE 2.2. Graph f.

Example 2.1.4.

Let $f : I \to \mathbb{R}$ be a C^r function where $I \subset \mathbb{R}$ is some open connected set. Then the graph of f is defined as follows:

$$\text{graph } f = \{\, (s,t) \in \mathbb{R}^2 \mid t = f(s), \quad s \in I \,\}. \tag{2.7}$$

Geometrically, graph f might appear as in Fig. 2.2.

We claim that graph f is a one-dimensional C^r manifold. In order to verify this we must show that the two requirements of Definition 2.1.3 can be satisfied.

1. Let $U = \mathbb{R}^2 \cap \text{graph } f$; then, by definition, $U = \text{graph } f$.

2. We define a coordinate chart on U in the following manner:

$$\begin{aligned} x : U &\to I, \\ \big(s, f(s)\big) &\mapsto s, \end{aligned} \tag{2.8}$$

with the inverse defined in the obvious way,

$$\begin{aligned} x^{-1} : I &\to U, \\ s &\mapsto \big(s, f(s)\big). \end{aligned} \tag{2.9}$$

It is clear that x and x^{-1} are C^r since f is C^r. Thus, graph f is a one-dimensional C^r manifold described by a single coordinate chart. We remark that this example should remind the reader of some of the heuristic aspects of elementary calculus, where it is common to visualize scalar functions as curves in the plane and to identify points on the curve with the corresponding points in the domain of the function.

End of Example 2.1.4

Example 2.1.5.
Consider the following set of points contained in $\mathbb{R}^3$:

$$M = \left\{ (u, v, w) \in \mathbb{R}^3 \mid u^2 + v^2 + w^2 = 1 \right\}. \qquad (2.10)$$

This is just the two-dimensional sphere of unit radius. We want to show that M is a two-dimensional C^∞ manifold.

Let us define the open sets on M:

$$
\begin{aligned}
U^1 &\equiv \left\{ (u, v, w) \in \mathbb{R}^3 \mid u^2 + v^2 + w^2 = 1,\ w > 0 \right\}, \\
U^2 &\equiv \left\{ (u, v, w) \in \mathbb{R}^3 \mid u^2 + v^2 + w^2 = 1,\ w < 0 \right\}, \\
U^3 &\equiv \left\{ (u, v, w) \in \mathbb{R}^3 \mid u^2 + v^2 + w^2 = 1,\ v > 0 \right\}, \\
U^4 &\equiv \left\{ (u, v, w) \in \mathbb{R}^3 \mid u^2 + v^2 + w^2 = 1,\ v < 0 \right\}, \\
U^5 &\equiv \left\{ (u, v, w) \in \mathbb{R}^3 \mid u^2 + v^2 + w^2 = 1,\ u > 0 \right\}, \\
U^6 &\equiv \left\{ (u, v, w) \in \mathbb{R}^3 \mid u^2 + v^2 + w^2 = 1,\ u < 0 \right\}. \qquad (2.11)
\end{aligned}
$$

It should be clear that these six sets are open with respect to M and that they cover M (see Fig. 2.3).

In these six sets, points of M can be represented as follows:

$$
\begin{aligned}
U^1 &: \quad \left(u, v, \sqrt{1 - u^2 - v^2} \right), \\
U^2 &: \quad \left(u, v, -\sqrt{1 - u^2 - v^2} \right), \\
U^3 &: \quad \left(u, \sqrt{1 - u^2 - w^2}, w \right), \\
U^4 &: \quad \left(u, -\sqrt{1 - u^2 - w^2}, w \right), \\
U^5 &: \quad \left(\sqrt{1 - v^2 - w^2}, v, w \right), \\
U^6 &: \quad \left(-\sqrt{1 - v^2 - w^2}, v, w \right). \qquad (2.12)
\end{aligned}
$$

We define maps of the U^α, $\alpha = 1, \ldots, 6$, into $\mathbb{R}^2$ as follows:

$$
\begin{aligned}
x^1 : \quad & U^1 \to B \subset \mathbb{R}^2, \\
& \left(u, v, \sqrt{1 - u^2 - v^2} \right) \mapsto (u, v),
\end{aligned}
$$

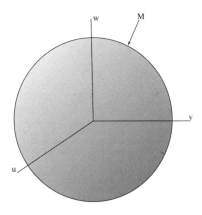

FIGURE 2.3. $M = U^1 \cup U^2 \cup U^3 \cup U^4 \cup U^5 \cup U^6$.

$$
\begin{aligned}
x^2: &\quad U^2 \to B \subset \mathbb{R}^2, \\
&\quad \left(u, v, -\sqrt{1 - u^2 - v^2}\right) \mapsto (u, v), \\
x^3: &\quad U^3 \to B \subset \mathbb{R}^2, \\
&\quad \left(u, \sqrt{1 - u^2 - w^2}, w\right) \mapsto (u, w), \\
x^4: &\quad U^4 \to B \subset \mathbb{R}^2, \\
&\quad \left(u, -\sqrt{1 - u^2 - w^2}, w\right) \mapsto (u, w), \\
x^5: &\quad U^5 \to B \subset \mathbb{R}^2, \\
&\quad \left(\sqrt{1 - v^2 - w^2}, v, w\right) \mapsto (v, w), \\
x^6: &\quad U^6 \to B \subset \mathbb{R}^2, \\
&\quad \left(-\sqrt{1 - v^2 - w^2}, v, w\right) \mapsto (v, w),
\end{aligned}
$$

$$(2.13)$$

with the inverse maps defined in the obvious manner (see Example 2.1.4) and where B is the open unit ball in $\mathbb{R}^2$. It should be clear that x^α and $(x^\alpha)^{-1}$, $\alpha = 1, \ldots, 6$, are C^∞.

Let us now demonstrate the compatibility of the coordinatizations on overlapping regions for a particular example. The open set in M,

$$U^1 \cap U^4 = \left\{ (u, v, w) \mid u^2 + v^2 + w^2 = 1,\ w > 0,\ v < 0 \right\}, \qquad (2.14)$$

may be given coordinates by either x^1 or x^4. The formulas for the coordinate changes are given as follows:

$$x^4 \circ \left(x^1\right)^{-1} : \quad x^1(U^1 \cap U^4) \to x^4(U^1 \cap U^4),$$
$$(u, v) \mapsto \left(u, \sqrt{1 - u^2 - v^2}\right) \equiv (u, w),$$
$$x^1 \circ \left(x^4\right)^{-1} : \quad x^4(U^1 \cap U^4) \to x^1(U^1 \cap U^4),$$
$$(u, w) \mapsto \left(u, -\sqrt{1 - u^2 - w^2}\right) \equiv (u, v). \quad (2.15)$$

It is easy to see that these two coordinate change maps are mutual inverses and that they are C^∞.

The reader should note the similarities between this example and Example 2.1.4. In the present example we were not able to represent the manifold globally as the graph of a function; however, we divided up the manifold into regions where we could represent it as a graph and, in these regions, the construction of the coordinate maps is exactly the same as in Example 2.1.4. Notice that the choice of (relatively) open sets to cover M is certainly not unique, but this does not result in any practical difficulties (see Spivak [1979] for a discussion of "maximal" atlases).

End of Example 2.1.5

2.2   Derivatives, Tangent Spaces, Normal Spaces, and Other "Structures" on Manifolds

The Tangent Space at a Point

In Definition 2.1.1 we considered the derivative of a map defined on a manifold. There is a geometric object associated with a manifold called the tangent space which plays an important role in the concept of the derivative of a function defined on a manifold. We want to motivate its construction by first recalling the definition of differentiability of a map defined on Euclidean space. We consider a map

$$f : U \to V, \quad (2.16)$$

where $U \subset \mathbb{R}^l$ and $V \subset \mathbb{R}^k$ are open sets. The map f is said to be *differentiable at a point* $x_0 \in U$ if there exists a linear map

$$L : \mathbb{R}^l \to \mathbb{R}^k, \quad (2.17)$$

such that for any $h \in \mathbb{R}^l$, $x_0 + h \in U$

$$|f(x_0 + h) - f(x_0) - Lh| = O\left(|h|^2\right), \quad (2.18)$$

where $|\cdot|$ is any norm on the appropriate Euclidean space. The linear map L is called the *derivative of f at x_0* and is represented by the $l \times k$ matrix of partial derivatives of f. Moreover, if L exists, it is unique. The linear map L acts on elements $h \in \mathbb{R}^l$ which can be viewed as vectors emanating from the point $x_0 \in U$. This previous sentence is quite important. The derivative is a linear map, but linearity of a map depends crucially on the linear structure of the space on which it operates. If we want to define the derivative of a map intrinsic to the manifold on which it is defined, we must somehow associate a linear space on which the derivative can operate in a way that is "natural" for the manifold. This linear space will be the tangent space at a point of the manifold at which the derivative is computed. We begin with two preliminary definitions.

Definition 2.2.1 *Let* $I = \{\, t \in \mathbb{R} \mid -\epsilon < t < \epsilon \,\}$ *for some fixed* $\epsilon > 0$. *Then a* C^r *curve in* M *is a* C^r *map* $C : I \to M$.

Definition 2.2.2 *Let* $C : I \to M$ *be a* C^1 *curve such that* $C(0) = p$. *Then the vector tangent to* C *at* p *is*

$$\frac{d}{dt}C(t)\Big|_{t=0} \equiv \dot{C}(0).$$

See Fig. 2.4 for an illustration of the geometry.

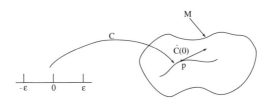

FIGURE 2.4. A Curve and Its Tangent Vector at a Point.

Before discussing the case of the tangent space at a point for a general manifold, let us first discuss the case where the manifold is $\mathbb{R}^m$ (see Example 2.1.3).

Let x be a point in $\mathbb{R}^m$; then define the tangent space to $\mathbb{R}^m$ at x, $T_x\mathbb{R}^m$, as the collection of tangent vectors to C^1 curves passing through x at x. It is easily seen that the set $T_x\mathbb{R}^m$ defined in this way can be identified with $\mathbb{R}^m$ itself. Indeed, every element $\xi \in \mathbb{R}^m$ can be thought of as a tangent vector of a C^1 curve, after shifting its "point of emanation" from $0 \in \mathbb{R}^m$ to $x \in \mathbb{R}^m$; e.g., take the curve $C(t) = x + t\xi$, $\xi \in \mathbb{R}^m$, then $dC(t)/dt|_{t=0} = \xi$, (see Fig. 2.5).

Recalling our brief discussion of the differentiation of maps defined on $\mathbb{R}^m$, it should now be clear what role the tangent space at a point plays in

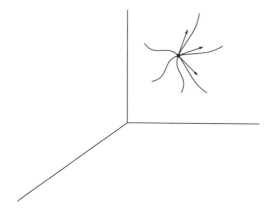

FIGURE 2.5. The Tangent Space of $\mathbb{R}^m$ at a Point $x \in \mathbb{R}^m$.

the definition of the derivative at a point; namely, $T_x\mathbb{R}^m$ is the domain of the derivative $Df(x)$, and locally it reflects the structure of the manifold $\mathbb{R}^m$, thus allowing the linear map $Df(x)$ to locally reflect the character of $f(x)$. In the case where the manifold has the structure of a linear vector space, we usually do not bother with formalizing the notion of the tangent space at a point since the tangent space at a point can be identified with the space itself. However, in the case where the domain of the map has no linear structure, in order to discuss a local linear approximation to the map at a point, i.e., the derivative of the map, it is necessary to introduce the structure of a linear vector space for the domain of the derivative since the definition of linearity of a map depends crucially on the fact that the domain of the map is linear.

Now we will define the tangent space at a point for an arbitrary differentiable manifold. Let $(U^\alpha; x^\alpha)$ be a chart containing the point p in the m-dimensional differentiable manifold M. Then $x^\alpha(U^\alpha)$ is an open set in $\mathbb{R}^m$ containing the point $x^\alpha(p)$. From the previous discussion, the tangent space at $x^\alpha(p)$ in $\mathbb{R}^m$, $T_{x^\alpha(p)}\mathbb{R}^m$, is just $\mathbb{R}^m$. To construct T_pM we carry $T_{x^\alpha(p)}\mathbb{R}^m$ to p in M by using x^α. Since $x^\alpha : U^\alpha \to \mathbb{R}^m$ is a diffeomorphism, $(x^\alpha)^{-1} : x^\alpha(U^\alpha) \to U^\alpha$ is also a diffeomorphism. Therefore, we can compute $D[(x^\alpha)^{-1}]$ which is a linear isomorphism mapping $\mathbb{R}^m$ into $\mathbb{R}^m$. The tangent space at $p \in M$, T_pM, is then defined to be $D[(x^\alpha)^{-1}]\big|_{x^\alpha(p)} \cdot \mathbb{R}^m$.

Definition 2.2.3 *Let $(U^\alpha; x^\alpha)$ be a chart on M with $p \in U^\alpha$. Then the tangent space to M at the point p, denoted T_pM, is defined to be $D[(x^\alpha)^{-1}]\big|_{x^\alpha(p)} \cdot \mathbb{R}^m$.*

The tangent space at p in M has the same geometrical interpretation as the tangent space at a point at $\mathbb{R}^m$; namely, it can be thought of as the collection of vectors tangent to curves which pass through p at p. This can be seen as follows: Let $(U^\alpha; x^\alpha)$ be a chart containing the point $p \in M$. Then, as previously discussed, $T_{x^\alpha(p)}\mathbb{R}^m$ consists of the collection of vectors tangent to differentiable curves passing through $x^\alpha(p)$ at $x^\alpha(p)$. Let $\gamma(t)$ be such a curve with $\gamma(0) = x^\alpha(p)$; then $\dot\gamma(0)$ is a vector tangent to $\gamma(t)$ at $x^\alpha(p)$. Using the fact that x^α is a diffeomorphism of U^α onto $x^\alpha(U^\alpha)$, $(x^\alpha)^{-1}(\gamma(t)) \equiv C(t)$ is a differentiable curve in M satisfying $C(0) = p$. Using the chain rule, the tangent vector to $C(t)$ at p is given by $D[(x^\alpha)^{-1}]\big|_{x^\alpha(p)} \cdot \dot\gamma(0) \equiv \dot C(0)$. Now $\dot\gamma(0)$ is a vector in $\mathbb{R}^m$; thus $\dot C(0)$ is an element of $T_p M$. So we see that the elements of $T_p M$ consist of the vectors tangent to differentiable curves passing through p at p. Finally, one must show the converse; namely, that *all* vectors tangent, at p, to differentiable curves in M passing through p are in $T_p M$. To show this, pick an arbitrary curve $C : I \to M$ such that $C(0) = p$. Since x^α is a C^r diffeomorphism, there is a C^r extension to an open set in $\mathbb{R}^n$ around p, $\bar x^\alpha$, and $\gamma(t) = \bar x^\alpha \circ C(t)$ is a curve in $\mathbb{R}^m$. (Note: It may be necessary to shrink the domain of definition of $\gamma(t)$.) By construction, $\gamma(0) = \bar x^\alpha(p)$ and, therefore, $\dot\gamma(0)$ is a vector in $T_{\bar x^\alpha(p)}(x^\alpha(U^\alpha)) = \mathbb{R}^m$. The equality $\dot C(0) = D[(x^\alpha)^{-1}]\big|_{x^\alpha(p)} \cdot \dot\gamma(0)$ then completes the argument.

Before leaving the tangent space at a point, there is one last detail to be considered; namely, in our construction of the tangent space at a point of a manifold we utilized a specific chart, but the tangent space is a geometrical object which should be intrinsic to the manifold, being representative of the manifold's local structure. Therefore, the tangent space should be independent of the specific chart used in its construction.

Proposition 2.2.1 *The construction of $T_p M$ is independent of the specific chart.*

Proof: Let $(U^\alpha; x^\alpha)$, $(U^\beta; x^\beta)$ be two charts with $U^\alpha \cap U^\beta \neq \emptyset$ and $p \in U^\alpha \cap U^\beta$. Then by Definition 2.2.2, $T_p M$ can be constructed as either $D[(x^\alpha)^{-1}]\big|_{x^\alpha(p)} \cdot \mathbb{R}^m$ or $D[(x^\beta)^{-1}]\big|_{x^\beta(p)} \cdot \mathbb{R}^m$. We must show that

$$D\left[(x^\alpha)^{-1}\right]\big|_{x^\alpha(p)} \cdot \mathbb{R}^m = D\left[(x^\beta)^{-1}\right]\big|_{x^\beta(p)} \cdot \mathbb{R}^m.$$

This can be established by the following argument. On $U^\alpha \cap U^\beta$ we have the relationship (see Fig. 2.1)

$$(x^\alpha)^{-1} = (x^\beta)^{-1} \circ \left[x^\beta \circ (x^\alpha)^{-1}\right]. \tag{2.19}$$

Differentiating (2.19) we get

$$D\left[(x^\alpha)^{-1}\right]\big|_{x^\alpha(p)} = D\left[(x^\beta)^{-1}\right]\big|_{x^\beta(p)} D\left[x^\beta \circ (x^\alpha)^{-1}\right]\big|_{x^\alpha(p)}, \tag{2.20}$$

but $D\left[x^\beta \circ (x^\alpha)^{-1}\right]$ is an isomorphism of $\mathbb{R}^m$ so $D\left[x^\beta \circ (x^\alpha)^{-1}\right]\big|_{x^\alpha(p)} \cdot$ $\mathbb{R}^m = \mathbb{R}^m$. Therefore, we get

$$
\begin{aligned}
D\left[(x^\alpha)^{-1}\right]\big|_{x^\alpha(p)} \cdot \mathbb{R}^m &= D\left[(x^\beta)^{-1}\right]\big|_{x^\beta(p)} D\left[x^\beta \circ (x^\alpha)^{-1}\right]\big|_{x^\alpha(p)} \cdot \mathbb{R}^m \\
&= D\left[(x^\beta)^{-1}\right]\big|_{x^\beta(p)} \cdot \mathbb{R}^m.
\end{aligned}
\tag{2.21}
$$

So we see that the tangent space at a point is independent of the particular chart chosen in a neighborhood of that point.   □

Now that we have defined the tangent space at a point of a manifold, we want to demonstrate the role it plays in defining the derivative of maps between manifolds. Let $f : M^m \to N^s$ be a C^r map where $M^m \subseteq \mathbb{R}^n$, $m \le n$, is an m-dimensional manifold and $N^s \subset \mathbb{R}^q$, $s \le q$, is an s-dimensional manifold. Let $(U^\alpha; x^\alpha)$ be a coordinate chart on M^m containing p and let $(W^\beta; y^\beta)$ be a coordinate chart on N^s containing $f(p)$. First, we must define what we mean by the *derivative of f*. Using the coordinate charts, we may write

$$
f(p) = (y^\beta)^{-1} \circ \left[y^\beta \circ f \circ (x^\alpha)^{-1}\right] \circ x^\alpha(p).
\tag{2.22}
$$

We then define the derivative of f at the point p as

$$
Df\big|_p = D\left[(y^\beta)^{-1}\right]\big|_{y^\beta(f(p))} D\left[y^\beta \circ f \circ (x^\alpha)^{-1}\right]\big|_{x^\alpha(p)} Dx^\alpha\big|_p.
\tag{2.23}
$$

One needs to show that this definition does not depend on the particular coordinate charts. However, this follows from an argument exactly like that used in Proposition 2.2.1.

Proposition 2.2.2 *Let $f : M^m \to N^s$ and Df be as defined above; then*

$$
Df\big|_p : T_p M^m \to T_{f(p)} N^s.
\tag{2.24}
$$

Proof: From (2.23) we have

$$
Df\big|_p D\left[(x^\alpha)^{-1}\right]\big|_{x^\alpha(p)} = D\left[(y^\beta)^{-1}\right]\big|_{y^\beta(f(p))} D\left[y^\beta \circ f \circ (x^\alpha)^{-1}\right]\big|_{x^\alpha(p)},
\tag{2.25}
$$

from which it follows that

$$
\begin{aligned}
Df\big|_p D\left[(x^\alpha)^{-1}\right]\big|_{x^\alpha(p)} \cdot \mathbb{R}^m \\
= D\left[(y^\beta)^{-1}\right]\big|_{y^\beta(f(p))} D\left[y^\beta \circ f \circ (x^\alpha)^{-1}\right]\big|_{x^\alpha(p)} \cdot \mathbb{R}^m.
\end{aligned}
\tag{2.26}
$$

However,

$$D\left[y^\beta \circ f \circ (x^\alpha)^{-1}\right]\big|_{x^\alpha(p)} \cdot \mathbb{R}^m \subset \mathbb{R}^s, \tag{2.27}$$

and

$$D\left[(y^\beta)^{-1}\right]\big|_{y^\beta(f(p))} \cdot \mathbb{R}^s = T_{f(p)}N^s. \tag{2.28}$$

So we see that $Df\big|_p \cdot T_p M^m \subset T_{f(p)}N^s$. □

Geometrically, Proposition 2.2.2 may be visualized as in Fig. 2.6.

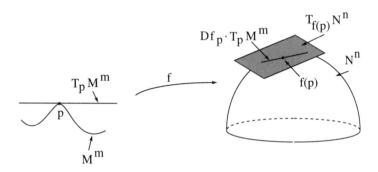

FIGURE 2.6. $T_p M^m$ and Its Image under $Df\big|_p$.

Our definition of the derivative of maps between manifolds also satisfies the *chain rule*. In particular, consider an additional map $g : N^s \to P^v$, where $P^v \subset \mathbb{R}^w$ is a C^r manifold ($v \le w$). Then we have the following proposition:

Proposition 2.2.3 $D\left(g \circ f\right)\big|_p = Dg\big|_{f(p)} \circ Df\big|_p$.

Proof: See Guillemin and Pollack [1974]. □

2.2.1   THE NORMAL SPACE AT A POINT

Let $\langle \cdot, \cdot \rangle$ denote the standard Euclidean inner product on $\mathbb{R}^n$ with $\| \cdot \|$ denoting the norm obtained from this inner product.

Definition 2.2.4 *For a point $p \in M \subset \mathbb{R}^n$ the set*

$$N_p = \{v \in \mathbb{R}^n \,|\, v \perp T_p M\} \tag{2.29}$$

is called the normal space at p.

In this definition the orthogonality symbol "$\perp$" is defined with respect to the Euclidean inner product. It should be clear that

$$T_p\mathbb{R}^n = T_pM + N_p,$$

where the $+$ sign denotes the direct sum of two vector spaces (the traditional sign of $\oplus$ for direct sums will be reserved for a different notion).

2.2.2   Tangent Bundles and Normal Bundles

When we study manifolds that are invariant under the flow generated by a vector field, it will be important to have information concerning the tangent spaces and normal spaces at different points on the manifold. In this regard, it is useful to consider the geometric object formed by the disjoint union of all the tangent spaces and normal spaces at all possible points of the manifold. These sets are called the *tangent bundle* and *normal bundle*, respectively. We first define the tangent bundle.

Definition 2.2.5 *The tangent bundle of a C^r manifold $M \subset \mathbb{R}^n$, denoted TM, is defined as*

$$TM = \{\, (p, v) \in M \times \mathbb{R}^n \mid v \in T_pM \,\}. \tag{2.30}$$

So the tangent bundle is the set of all possible ordered pairs of points of M and tangent vectors to M at the same point. TM itself has the structure of a $2m$-dimensional C^{r-1} manifold, as we show next.

Proposition 2.2.4 *Let $M \subset \mathbb{R}^n$ be a C^r manifold of dimension m; then the tangent bundle of M, $TM \subset \mathbb{R}^{2n}$, is a C^{r-1} manifold of dimension $2m$.*

Proof: We must construct an atlas for TM. Let $(U^\alpha; x^\alpha)$, $\alpha \in A$, be an atlas for M. Then $(U^\alpha, TM|_{U^\alpha}; x^\alpha, Dx^\alpha)$ is an atlas for TM, which is C^{r-1} since Dx^α is C^{r-1}. For the remaining details of the proof, see Guillemin and Pollack [1974] or Berger and Gostiaux [1988].   □

Next we define the *normal bundle*.

Definition 2.2.6 *The set*

$$N = \{(p, v) \in M \times \mathbb{R}^n \mid v \perp T_pM\} \tag{2.31}$$

is referred to as the normal bundle of M.

Like the tangent bundle, the normal bundle is also a C^{r-1} manifold.
    By construction, we have for any $p \in M$

$$T_p\mathbb{R}^n = \mathbb{R}^n = T_pM + N_p, \tag{2.32}$$

where $+$ denotes the vector space sum. The map

$$\Pi_p : T_p\mathbb{R}^n \to N_p \tag{2.33}$$

will denote the standard orthogonal projection map, using the standard inner product on $\mathbb{R}^n$, that is familiar from linear algebra.

For bundles this notion generalizes as follows:

$$T\mathbb{R}^n|_M = TM \oplus N, \tag{2.34}$$

where the symbol $\oplus$ refers to the *Whitney sum* of the tangent and normal bundles. This can be thought of as taking the disjoint union of (2.32). The map

$$\Pi : T\mathbb{R}^n|_M \to N \tag{2.35}$$

denotes the orthogonal bundle projection map that is defined pointwise using (2.33).

VECTOR BUNDLES

The tangent and normal bundles of a manifold are examples of *vector bundles*. Roughly speaking, a vector bundle is a manifold that has a linear vector space associated to each point. For the tangent and normal bundles of a manifold the vector spaces at each point $p \in M$ are T_pM and N_p, respectively. These vector spaces are isomorphic to $\mathbb{R}^m$ and $\mathbb{R}^{n-m}$, respectively. The individual vector spaces, N_p, that make up the normal bundle N are referred to as the *fibers* of the vector bundle and p is referred to as the *basepoint* of the fiber N_p. The same terminology applies to TM and T_pM. The *fiber projection map*, denoted ω, is the map that maps any point in a fiber to the basepoint of the fiber. More, precisely the fiber projection map is defined as follows:

$$
\begin{aligned}
\omega : N &\to M, \\
(p, v) &\to p.
\end{aligned}
\tag{2.36}
$$

By a *section* of the normal bundle, denoted u, we mean a map that takes points in M to points in N, i.e.,

$$
\begin{aligned}
u : M &\to N, \\
p &\to (p, v).
\end{aligned}
\tag{2.37}
$$

We say that u is a C^r *section* if it is a C^r map in the sense described in Definitions 2.1.1 and 2.1.2. The *zero section* of M is $M \times \{0\}$ and has the obvious identification with M. Clearly, the composition $\omega \circ u$ is the identity map on M.

Vector bundles, in particular the normal bundle, have the important property of *local triviality*. This means that the normal bundle, restricted to a sufficiently small neighborhood in M, say $U \subset M$, is C^r diffeomorphic to a cartesian product of U with $\mathbb{R}^{n-m}$. More precisely, for every point $p \in M$ there is a neighborhood $U_p \subset M$ and a C^r diffeomorphism $\alpha : N|_{U_p} \to U_p \times \mathbb{R}^{n-m}$ that restricts to a linear isomorphism $N_{p'} \to \{p'\} \times \mathbb{R}^{n-m}$ for every $p' \in U_p$. An important consequence of this property is that a basis of the linear vector space N_p can be chosen to vary in a C^r manner with respect to the basepoint p' *for all $p' \in U_p$*.

2.2.3 THE NEIGHBORHOOD OF A MANIFOLD

We now want to show how the normal bundle can be used to construct a neighborhood of a manifold. Consider the map

$$
\begin{aligned}
h : N &\to \mathbb{R}^n, \\
(p, v) &\mapsto p + v.
\end{aligned}
\tag{2.38}
$$

This map provides a natural identification of $N|_M$ with $\mathbb{R}^n$. Consider the following *subset* of the normal bundle

$$
N_\epsilon = \{(p, v) \in N \,|\, |v| < \epsilon\}.
\tag{2.39}
$$

We have the following result.

Proposition 2.2.5 *Let $K \subset M$ be a compact set. Then there exists $\epsilon_0 > 0$ such that for $0 < \epsilon < \epsilon_0$, h maps $N_\epsilon|_K$ C^{r-1} diffeomorphically onto a neighborhood of K in $\mathbb{R}^n$.*

Proof: We will use the following "global" version of the inverse function theorem.

Lemma 2.2.6 *Suppose X and Y are C^r manifolds of the same dimension and let $f : X \to Y$ be a C^r map that is one-to-one on a compact C^r submanifold Z of X (i.e., Z is a subset of X which also has the structure of a C^r manifold). Suppose that for all $x \in Z$*

$$
Df|_x : T_x X \to T_{f(x)} Y
$$

is an isomorphism. Then f maps an open neighborhood of Z in X C^r diffeomorphically onto an open neighborhood of $f(Z)$ in Y.

Proof: This is exercise 10 in chapter 1, section 3 of Guillemin and Pollack [1974]. Using partitions of unity it can also be extended to the case for noncompact X and Z (exercise 14, chapter 1, section 8 of Guillemin and Pollack [1974]).   □

To prove the proposition we need to show that h is one-to-one on $K \times \{0\}$ and that $Dh|_x$ is an isomorphism for each $x \equiv (p, 0) \in K \times \{0\} \subset N_\epsilon|_K$. The result then follows from Lemma 2.2.6.

It suffices to prove the result locally, and following this, we will also use the local triviality of the normal bundle. Let $(U; x)$ be a chart with $U \subset K$. Then, by the local triviality of the normal bundle, we can choose U smaller if necessary, so that $N|_U$ is C^{r-1} diffeomorphic to $U \times \mathbb{R}^{n-m}$. Thus, we can regard h as being defined on $U \times \mathbb{R}^{n-m}$. We denote a point $(p, v) \in U \times \mathbb{R}^{n-m}$ by $(p, v_1, \ldots, v_{n-m})$ where the components of v are with respect to the standard basis of $\mathbb{R}^{n-m}$. Then the following map is a local coordinate representation for the map h:

$$\psi \equiv x \times id : \mathbb{R}^m \times \mathbb{R}^{n-m} \quad \rightarrow \quad \mathbb{R}^n,$$
$$(x(p), v) \quad \mapsto \quad (x_1, \ldots, x_m, v_1, \ldots, v_{n-m}),$$

where id denote the identity map on $\mathbb{R}^{n-m}$. Obviously, when restricted to $K \times \{0\}$, ψ is one-to-one. A simple calculation shows that

$$D\psi = \begin{pmatrix} Dx & 0 \\ 0 & id \end{pmatrix}. \tag{2.40}$$

So $D\psi$ is an isomorphism. This completes the proof of the result.   □

Example 2.2.1.

We now give an example that illustrates all of the concepts developed thus far, and allows for explicit calculations. Consider the two-sphere in $\mathbb{R}^3$, $S^2 \subset \mathbb{R}^3$, given by

$$S^2 = \left\{ (x, y, z) \in \mathbb{R}^3 \mid x^2 + y^2 + z^2 = 1 \right\}.$$

In Example 2.1.5 we showed that S^2 is a two-dimensional C^∞ manifold.

Another way of viewing S^2 is that it is given by the zero set of the function

$$f(x, y, z) \equiv x^2 + y^2 + z^2 - 1.$$

Let $C(t) = (x(t), y(t), z(t))$, $t \in \mathcal{I} \subset \mathbb{R}$, denote a curve on S^2 (where $\mathcal{I}$ is some interval which, without loss of generality, contains the origin) and let $C(0) \equiv p = (x_0, y_0, z_0)$. Then

$$\frac{df}{dt}(C(0)) = \langle Df(C(0)), \dot{C}(0) \rangle = 0$$
$$= \langle (2x_0, 2y_0, 2z_0), (\dot{x}(0), \dot{y}(0), \dot{z}(0)) \rangle = 0. \tag{2.41}$$

Since $\dot{C}(0)$ is a vector tangent to S^2 at $C(0) = p$, it follows from this calculation that

$$T_p S^2 \simeq \left\{ (a, b, c) \in \mathbb{R}^3 \,|\, ax_0 + by_0 + cz_0 = 0 \right\}.$$

From this expression we see that $T_p S^2$ is a two-dimensional linear vector space (a plane) passing through the origin of $\mathbb{R}^3$. Moreover, we immediately have

$$N_p = \left\{ (a, b, c) \in \mathbb{R}^3 \,|\, (a, b, c) = \alpha(x_0, y_0, z_0), \; \forall \alpha \in \mathbb{R} \right\}.$$

From this expression we see that N_p is a one-dimensional linear vector space (a line) passing through the origin in $\mathbb{R}^3$. By construction we see that

$$\mathbb{R}^3 = T_p S^2 + N_p.$$

With expressions for the tangent and normal spaces at a point in hand, we can then compute explicit expressions for the tangent and normal bundles.

The tangent bundle of S^2 is the disjoint union of the tangent spaces at all points of S^2. Hence, we have

$$TS^2 = \big\{ (p, v) \,|\, p = (x_0, y_0, z_0) \in \mathbb{R}^3 \quad \text{and} \quad x_0^2 + y_0^2 + z_0^2 = 1,$$

$$v = (v_1, v_2, v_3) \in \mathbb{R}^3, \text{ and } v_1 x_0 + v_2 y_0 + v_3 z_0 = 0 \big\}.$$

Clearly, TS^2 is four dimensional. In the same manner, the normal bundle is given by

$$N = \big\{ (p, v) \,|\, p = (x_0, y_0, z_0) \in \mathbb{R}^3 \quad \text{and} \quad x_0^2 + y_0^2 + z_0^2 = 1,$$

$$v = (v_1, v_2, v_3) \in \mathbb{R}^3 \text{ and } (v_1, v_2, v_3) = \alpha(x_0, y_0, z_0), \alpha \in \mathbb{R} \big\},$$

and it is three dimensional.

The tangent bundle of $\mathbb{R}^3$, restricted to S^2 (meaning all tangent vectors emanate from points on S^2) is given by

$$
\begin{aligned}
T\mathbb{R}^3|_{S^2} &= \big\{ (p, v) \,|\, p \in S^2, \, v \in \mathbb{R}^3 \big\} \\
&= \big\{ (p = (x, y, z), v) \,|\, x^2 + y^2 + z^2 = 1, v \in \mathbb{R}^3 \big\}.
\end{aligned}
$$

We can immediately verify that

$$T\mathbb{R}^3|_{S^2} = TS^2 \oplus N.$$

In other words, at each point $p \in S^2$, $T_p\mathbb{R}^3 = T_p S^2 + N_p$, where the $+$ in this term is the usual direct sum of vector spaces.

Now we construct the orthogonal projection map for this example:

$$\Pi : T\mathbb{R}^3|_{S^2} \to N.$$

Let $(p, v) \in T\mathbb{R}^3|_{S^2}$; then we have

$$\Pi : (p, v) \mapsto (p, \langle v, n \rangle n),$$

where n is a unit vector normal to S^2 at p, i.e., $n = Df(p)/\parallel Df(p) \parallel$. If we take $p = (x_0, y_0, z_0)$, then we can explicitly compute

$$n = \left(\frac{x_0}{\sqrt{x_0^2 + y_0^2 + z_0^2}}, \frac{y_0}{\sqrt{x_0^2 + y_0^2 + z_0^2}}, \frac{z_0}{\sqrt{x_0^2 + y_0^2 + z_0^2}} \right);$$

however, $x_0^2 + y_0^2 + z_0^2 = 1$, so we have

$$n = (x_0, y_0, z_0).$$

Finally, we use Proposition 2.2.5 to construct a neighborhood of S^2. This involves a careful consideration of the following map:

$$\begin{aligned} h : N &\to \mathbb{R}^3, \\ (p, v) &\mapsto p + v. \end{aligned}$$

Consider a point $(p, v) \in N$ and let $p = (x_0, y_0, z_0)$. Then $v \in N_p$ is of the form $v = \alpha n$, where n is a vector normal to S^2 at p and $\alpha \in \mathbb{R}$. However, from the above calculation we have $n = (x_0, y_0, z_0)$. Hence, we find the map h of the form

$$\begin{aligned} h : N &\to \mathbb{R}^3, \\ (p, v) &\mapsto p + v = (1 + \alpha)(x_0, y_0, z_0), \end{aligned}$$

where α is the coordinate on the fibers. Thus, varying α and x_0, y_0, and z_0, subject to $x_0^2 + y_0^2 + z_0^2 = 1$, sweeps out a neighborhood of S^2 in $\mathbb{R}^3$.

_____
End of Example 2.2.1

2.3   Manifolds with Boundary

Before proceeding to discuss invariant manifolds of ordinary differential equations, we need to discuss the idea of a manifold with boundary. Note that in the definition of a differentiable manifold (Definition 2.1.3) each point of the manifold has a neighborhood diffeomorphic to some open set in

$\mathbb{R}^m$. This rules out the possibility of boundary points. However, manifolds with boundary arise frequently in applications, and we now want to give a definition of a C^r manifold with boundary. We begin with a preliminary definition.

Definition 2.3.1 *The closed half-space, $H^m \subset \mathbb{R}^m$, is defined as follows*

$$H^m = \{\,(x_1, x_2, \ldots, x_m) \in \mathbb{R}^m \mid x_1 \geq 0\,\}. \tag{2.42}$$

The boundary of H^m, denoted ∂H^m, is $\mathbb{R}^{m-1}$.

We now give the definition of a differentiable manifold with boundary.

Definition 2.3.2 *A subset $M \subset \mathbb{R}^n$ is called a C^r manifold of dimension m with boundary if it possesses the following two structural characteristics:*

1. *There exists a countable collection of open sets $V^\alpha \subset \mathbb{R}^n$, $\alpha \in A$ where A is some countable index set, with $U^\alpha \equiv V^\alpha \cap M$ such that $M = \bigcup_{\alpha \in A} U^\alpha$.*

2. *There exists a C^r diffeomorphism x^α defined on each U^α which maps U^α onto some set $W \cap H^m$ where W is some open set in $\mathbb{R}^n$.*

We make the following remarks concerning Definition 2.3.2.

1. The *boundary of M*, denoted ∂M, is defined to be the set of points in M which are mapped to ∂H^m under x^α. It is necessary to show that this set is independent of the particular chart that is chosen; see Guillemin and Pollack [1974] for the details.

2. The boundary of M is a C^r manifold of dimension $m-1$, and $M - \partial M$ is a C^r manifold of dimension m.

3. The tangent space of M at a point is defined just as in Definition 2.2.3 even if the point is a boundary point.

Example 2.3.1.
    Consider the finite cylinder in $\mathbb{R}^3$ defined by

$$\mathcal{C} = \{\,(u, v, w) \in \mathbb{R}^3 \mid u^2 + v^2 = 1, 0 \leq w \leq 1\,\}. \tag{2.43}$$

We will show that $\mathcal{C}$ is a C^∞ manifold with boundary. The following sets form a covering of $\mathcal{C}$:

$$U^1 = \{(u, v) \,|\, u^2 + v^2 = 1, (u, v) \neq (1, 0)\} \times \{w \,|\, 0 \leq w < 1\},$$

$$U^2 = \{(u, v) \,|\, u^2 + v^2 = 1, (u, v) \neq (-1, 0)\} \times \{w \,|\, 0 \leq w < 1\},$$

$$\tag{2.44}$$

$$U^3 = \{(u, v) \,|\, u^2 + v^2 = 1, (u, v) \neq (1, 0)\} \times \{w \,|\, 0 < w \leq 1\},$$

$$U^4 = \{(u, v) \,|\, u^2 + v^2 = 1, (u, v) \neq (-1, 0)\} \times \{w \,|\, 0 < w \leq 1\}.$$

On each set we define diffeomorphisms with H^2 via sterographic projections as follows:

$$x^1 : (u, v, w) \mapsto \left(\tfrac{v}{1-|u|}, w\right),$$

$$x^2 : (u, v, w) \mapsto \left(\tfrac{v}{1-|u|}, w\right),$$

$$\tag{2.45}$$

$$x^3 : (u, v, w) \mapsto \left(\tfrac{v}{1-|u|}, 1 - w\right),$$

$$x^4 : (u, v, w) \mapsto \left(\tfrac{v}{1-|u|}, 1 - w\right).$$

End of Example 2.3.1

2.4 An Example

We end this background section with an example that is central to the development of the global perturbation methods found in Wiggins [1988]. Moreover, it will serve to illustrate all of the concepts developed in this section.

Consider the following system of ordinary differential equations:

$$\begin{aligned}
\dot{x} &= JD_x H(x, I), & \tag{2.46}\\
\dot{I} &= 0, & \tag{2.47}\\
\dot{\theta} &= \Omega(x, I), & \tag{2.48}
\end{aligned}$$

with

$$(x, I) \in V \subset \mathbb{R}^{2n} \times \mathbb{R}^m, \ \theta \in T^l, \ n, l \geq 1, \ m \geq 0,$$

and assume that $H : V \to \mathbb{R} \in C^{r+1}$, $\Omega \in C^r$. J is the standard $2n \times 2n$ symplectic matrix given by

$$\begin{pmatrix} 0 & id^{n \times n} \\ -id^{n \times n} & 0 \end{pmatrix}$$

where $id^{n \times n}$ denotes the $n \times n$ identity matrix.

Assumption: For every $I \in U \subset \mathbb{R}^m$, (2.46) has a hyperbolic fixed point denoted by $x = \gamma(I)$.

It follows from this assumption that the set

$$\mathcal{M} = \{(x, I, \theta) \subset V \times T^l \,|\, x = \gamma(I), \, I \in U\} \tag{2.49}$$

is an *invariant set* under the dynamics of this system of ordinary differential equations; i.e., the vector field is tangent to this set. We will illustrate the ideas developed in this section by exploring the geometry of this invariant set further.

$\boxed{\mathcal{M} \text{ is a } C^r \text{ Manifold.}}$

From the assumption, we have

$$D_x H(\gamma(I), I) = 0, \tag{2.50}$$

for all $I \in U$. Differentiating both sides of (2.50) with respect to I yields

$$D_I \gamma(I) = - \left[D_x^2 H(\gamma(I), I) \right]^{-1} D_I D_x H(\gamma(I), I). \tag{2.51}$$

(Note that the existence of $\left[D_x^2 H(\gamma(I), I) \right]^{-1}$ follows from the assumption that $x = \gamma(I)$ is a hyperbolic fixed point.) Since the right-hand side of (2.51) is a C^{r-1} function of I, we have that $\gamma : U \to \mathbb{R}^{2n} \times \mathbb{R}^m \times T^l$ is a C^r function. From this it follows that $\mathcal{M}$ is the graph of the C^r function

$$\begin{aligned} f : U \times T^l &\to \mathbb{R}^{2n}, \\ (I, \theta) &\mapsto \gamma(I) \equiv f(I, \theta); \end{aligned} \tag{2.52}$$

hence, by the same argument given in Example 2.1.4, it follows that $\mathcal{M}$ is a C^r manifold of dimension $m + l$.

For the following, we locally embed $\mathcal{M}$ in $\mathbb{R}^{2n+m+2l} \equiv \mathcal{P}$ with the following map:

$$\begin{aligned} F : U \times T^l &\to \mathcal{P}, \\ (I, \theta) &\to (\gamma(I), I, s(\theta)), \end{aligned} \tag{2.53}$$

where

$$s : T^l \rightarrow \mathbb{R}^{2l},$$
$$(\theta_1, \cdots, \theta_l) \mapsto (\cos\theta_1, \sin\theta_1, \ldots, \cos\theta_l, \sin\theta_l), \qquad (2.54)$$

and we will consider $\mathcal{M} = F(U \times T^l) \subset \mathcal{P}$. Note that the maps F and s are C^r diffeomorphisms in the sense of Definitions 2.1.1 and 2.1.2.

The Construction of $T\mathcal{P}|_{\mathcal{M}}$.

Since for any $p \in \mathcal{P}$, $T_p\mathcal{P}$ is isomorphic to $\mathbb{R}^{2n+m+2l}$, we can immediately write down

$$T\mathcal{P} = \left\{ (x, I, \eta, u, v, w) \mid x, u \in \mathbb{R}^{2n}, \ I, v \in \mathbb{R}^m, \ \eta, w \in \mathbb{R}^{2l} \right\}. \qquad (2.55)$$

Using (2.49) and (2.55), we obtain

$$T\mathcal{P}|_{\mathcal{M}} = \big\{ (x, I, \eta, u, v, w) \mid x = \gamma(I), \ \eta = s(\theta), \ I \in U, \ \theta \in T^l,$$
$$(u, v, w) \in \mathbb{R}^{2n+m+2l} \big\}, \qquad (2.56)$$

which shows that $T\mathcal{P}|_{\mathcal{M}}$ is a C^{r-1} manifold of dimension $m + l + (2n + m + 2l) = 2n + 2m + 3l$.

The Construction of the Splitting of $T\mathcal{P}|_{\mathcal{M}}$ into $T\mathcal{M}$ and N.

---

1. First we construct $T\mathcal{M}$. Let $p \in \mathcal{M}$ and $W \subset \mathcal{M}$ be an open set in $\mathcal{M}$ with $p \in W$. Then from (2.53) it follows that $(W; G)$ is a chart for $\mathcal{M}$ around p where

$$G : W \rightarrow \mathbb{R}^{m+l},$$
$$(x, I, \eta) \mapsto F^{-1}(x, I, \eta) = (I, s^{-1}(\eta)). \qquad (2.57)$$

(If necessary, we must shrink W so that $\theta = s^{-1}(\eta) \in [0, 2\pi) \times \cdots \times [0, 2\pi) \subset \mathbb{R}^l$, i.e. so that s^{-1} is a diffeomorphism.) Using Definition 2.2.3, we calculate $T_p\mathcal{M}$ by

$$T_p\mathcal{M} = D\left[G^{-1}\right]_{G(p)} \cdot \mathbb{R}^{m+l} = DF|_{(I,\theta)} \cdot \mathbb{R}^{m+l} = \text{Range}(A(I, \theta)), \qquad (2.58)$$

where $A(I, \theta) \in \mathbb{R}^{(2n+m+2l) \times (m+l)}$ is the matrix representation of $DF|_{(I,\theta)}$ with respect to the standard basis $\{e_j\}_{j=1}^{m+l}$ in $\mathbb{R}^{m+l}$ and the standard basis $\{f_k\}_{k=1}^{2n+m+2l}$ in $T_p\mathcal{P} = \mathbb{R}^{2n+m+2l}$. By assumption, $A(I,\theta)$ is C^{r-1} in its arguments. Using (2.53) we calculate

$$A(I,\theta) = \begin{pmatrix} D_I\gamma(I) & 0^{2n\times l} \\ \\ id^{m\times m} & 0^{m\times l} \\ \\ 0^{2l\times m} & T(\theta) \end{pmatrix}, \qquad (2.59)$$

where $D_I\gamma \in \mathbb{R}^{2n\times m}$ and $T \in \mathbb{R}^{2l\times l}$ with

$$T(\theta) = \begin{pmatrix} -\sin\theta_1 & 0 & \cdots & 0 \\ \cos\theta_1 & 0 & \cdots & 0 \\ 0 & -\sin\theta_2 & \cdots & 0 \\ 0 & \cos\theta_2 & \cdots & 0 \\ 0 & 0 & \cdots & 0 \\ \vdots & \vdots & & \vdots \\ 0 & 0 & \cdots & -\sin\theta_l \\ 0 & 0 & \cdots & \cos\theta_l \end{pmatrix}. \qquad (2.60)$$

It is easy to see from (2.59) and (2.60) that all the columns of $A(I,\theta)$ are linearly independent; therefore, they span $\mathrm{Range}(A(I,\theta))$. We then define the vectors

$$c'_j(I,\theta) = \frac{c_j(I,\theta)}{\|c_j(I,\theta)\|} \in \mathbb{R}^{2n+m+2l}, \qquad j = 1,\ldots,m+l, \qquad (2.61)$$

where c_j is the jth column of A, and $\|\cdot\|$ denotes the Euclidean norm on $\mathbb{R}^{2n+m+2l}$. Then, using (2.58), we can write

$$T\mathcal{M} = \{(x,I,\eta,u,v,w)|x=\gamma(I),\ \eta=s(\theta),\ (u,v,w)$$

$$= \Sigma_{j=1}^{m+l}\lambda_j c'_j(I,\theta) \in \mathbb{R}^{2n+m+2l},$$

$$I \in U,\ \theta \in T^l,\ \lambda_j \in \mathbb{R},\ j = 1,\ldots,m+l\}.$$
$$(2.62)$$

(A local parametrization of $T\mathcal{M}$ can be obtained by taking θ from a sufficiently small open subset of $\mathbb{R}^m$, instead of T^l.)

2. Next we construct the normal bundle, N. Recall that for any point $p \in \mathcal{M}$, the corresponding fiber of N is defined as

$$N_p = \{(p, z) \in \mathcal{M} \times T_p\mathcal{P} \mid z \perp T_p\mathcal{M}\}. \tag{2.63}$$

Hence, to construct N_p we need to find the orthogonal complement, $T_p\mathcal{M}^\perp$, of $T_p\mathcal{M}$ within $T_p\mathcal{P}$ with respect to the standard metric on $T_p\mathcal{P}$. To make the calculation simpler, we can use the Euclidean structure of $T_p\mathcal{P}$ which it inherits from the natural identification of $\mathcal{P}$ with $\mathbb{R}^{2n+m+2l}$. In other words, we may use the standard inner product $\langle \cdot, \cdot \rangle \colon \mathcal{P} \times \mathcal{P} \to \mathbb{R}$ to construct $T_p\mathcal{M}^\perp$. Technically speaking, this amounts to finding the orthogonal complement of span $(c_1, \ldots, c_{m+l})$ or finding $\ker(A^t(I, \theta))$ with $A(I, \theta)$ defined in (2.59). Let $z = (z_{2n}, z_m, z_{2l}) \in \mathbb{R}^{2n+m+2l}$. Then from (2.59) and (2.60) we see that

$$A^t(I, \theta)z = 0 \tag{2.64}$$

is equivalent to

$$\begin{aligned} z_m &= -\left[D_I\gamma(I)\right]^t z_{2n}, \\ T^t(\theta)z_{2l} &= 0 \in \mathbb{R}^{2l}. \end{aligned} \tag{2.65}$$

Hence, a solution of (2.64) can be constructed as follows: Select $2n$ linearly independent $2n$-vectors and substitute them for z_{2n} in (2.65), respectively, to obtain the corresponding m-vectors for z_m. Also, set $z_{2l} = 0$. This way we obtain $2n$ linearly independent solutions of (2.64). We can obtain l more linearly independent solutions of the second equation of (2.65) for z_{2l}. Carrying out this procedure, we obtain

$$N_p \simeq \ker(A^t(I, \theta)) = \text{span}\,(d_1(I, \theta), \ldots, d_{2n+l}(I, \theta)), \tag{2.66}$$

with

$$
d_k(I,\theta) =
\begin{pmatrix}
\left.\begin{matrix} 0 \\ \vdots \\ 1 \\ \vdots \\ 0 \end{matrix}\right\} 2n \\
\left.-\left(D_I\gamma_k(I)\right)^t\right\} m \\
\left.\begin{matrix} 0 \\ \vdots \\ 0 \end{matrix}\right\} 2l
\end{pmatrix}
\qquad \text{for} \quad k = 1,\ldots,2n \quad (2.67)
$$

$$
d_k(I,\theta) =
\begin{pmatrix}
\left.\begin{matrix} 0 \\ \vdots \\ 0 \\ 0 \end{matrix}\right\} 2n+m \\
\left.\begin{matrix} 0 \\ \vdots \\ \cos\theta_{k-2n} \\ \sin\theta_{k-2n} \\ \vdots \\ 0 \end{matrix}\right\} 2l
\end{pmatrix}
\qquad \text{for } k = 2n+1,\ldots,2n+l.
$$

$$(2.68)$$

The right braces denote different blocks in the column vectors and the numbers on the immediate right indicate the number of entries in each block. In (2.67) the 1 in the first block of length $2n$ occurs in the kth place. In (2.68), $\sin\theta_{k-2n}$ and $\cos\theta_{k-2n}$ appear in the $(2k+m-2n)^{\text{th}}$ and $(2k+m-2n-1)^{\text{th}}$ places in the column vector, respectively. For later purposes we normalize the vectors d_k by letting

$$
d'_k(I,\theta) = \frac{d_k(I,\theta)}{\|\,d_k(I,\theta)\,\|}, \qquad k = 1,\ldots,2n+l. \qquad (2.69)
$$

Then using (2.63), (2.66), and (2.69) we can write

$$N = \{(x, I, \eta, u, v, w) \mid x = \gamma(I), \, \eta = s(\theta), \, (u, v, w)$$

$$= \Sigma_{k=1}^{2n+l} \lambda_k d_k'(I, \theta) \in \mathbb{R}^{2n+m+2l},$$

$$I \in U, \, \theta \in T^l, \lambda_k \in \mathbb{R}, \, k = 1, \ldots, 2n+l\}.$$
$$(2.70)$$

(Again, for an actual local parametrization of N we have to restrict θ to a small enough open subset of $\mathbb{R}^l$.)

By construction, we now have

$$T_p\mathcal{P} = T_p\mathcal{M} + N_p, \qquad (2.71)$$

where $+$ denotes the direct sum of vector spaces, and we also have

$$T\mathcal{P}|_{\mathcal{M}} = T\mathcal{M} \oplus N, \qquad (2.72)$$

where $\oplus$ denotes the Whitney sum of vector bundles.

Construction of the Orthogonal Projection Map $\Pi : T\mathcal{P}|_{\mathcal{M}} \to N$.

We will construct the projection map for a single fiber with basepoint $p \in \mathcal{M}$; i.e., for some $p = (x, I, \theta) \in \mathcal{M}$, we calculate

$$\Pi_p : T_p\mathcal{P} \to N_p. \qquad (2.73)$$

Fix the standard basis $\{e_j\}_{j=1}^{2n+m+2l}$ for $T_p\mathcal{P} = \mathbb{R}^{2n+m+2l}$ and the basis $\left\{d_k'(I, s^{-1}(\eta))\right\}_{k=1}^{2n+l}$ in N_p. We can now compute the matrix representation $B(I, \theta) \in \mathbb{R}^{(2n+m+2l)\times(2n+l)}$ of the linear map Π_p defined in (2.73). Clearly,

$$\Pi_p(e_j) = \Sigma_{k=1}^{2n+l} \langle e_j, d_k'(I, \theta)\rangle d_k'(I, \theta) = \Sigma_{k=1}^{2n+l} B_{jk}(I, \theta) d_k'(I, \theta). \qquad (2.74)$$

Using (2.68), (2.69), and (2.74) yields

$$
B_{jk}(I,\theta) = \begin{cases}
\dfrac{\delta_{j,k}}{\sqrt{1+\parallel D_I\gamma_k(I)\parallel^2}} & \text{if} \quad 1\le k, j \le 2n, \\[4mm]
0 & \text{if} \quad 1\le j \le 2n, \\
& \qquad 2n < k \le 2n+l, \\[3mm]
\dfrac{-D_{I_{j-2n}}\gamma_k(I)}{\sqrt{1+\parallel D_I\gamma_k(I)\parallel^2}} & \text{if} \quad 2n < j \le 2n+m, \\[2mm]
& \qquad 1\le k \le 2n, \\[3mm]
0 & \text{if} \quad 2n < j \le 2n+m, \\
& \qquad 2n < k \le 2n+l, \\[3mm]
0 & \text{if} \quad 2n+m < j \le 2n+m+2l, \\
& \qquad 1\le k \le 2n, \\[3mm]
\delta_{j,2(k-n)+m-1}\cos\theta_{k-2n} & \text{if} \quad 2n+m < j \le 2n+m+2l, \\
& \qquad 2n < k \le 2n+l, \\
& \qquad k-2n \text{ odd}, \\[3mm]
\delta_{j,2(k-n)+m}\sin\theta_{k-2n} & \text{if} \quad 2n+m < j \le 2n+m+2l, \\
& \qquad 2n < k \le 2n+l, \\
& \qquad k-2n \text{ even}, \\
\end{cases}
\tag{2.75}
$$

where $\delta_{j,k}$ denotes the Kronecker delta. Having a representation of Π_p, we can now write down Π as

$$
\begin{aligned}
\Pi : TP|_{\mathcal{M}} &\rightarrow N, \\
(x,I,\eta,z) &\mapsto (x,I,\eta,B(I,s^{-1}(\eta))z),
\end{aligned}
\tag{2.76}
$$

with B defined in (2.75) and $z \in \mathbb{R}^{2n+m+2l}$.

Construction of a Neighborhood of $\mathcal{M}$ via the Map $h : N \rightarrow \mathcal{P}$.

Recall the map

$$
\begin{aligned}
h : N &\rightarrow \mathcal{P}, \\
(p,v) &\mapsto p+v
\end{aligned}
\tag{2.77}
$$

introduced in (2.38) and described through Proposition 2.2.5. We now want to give an explicit construction of this map for our example. In our coor-

2.4. An Example   49

dinates, with N defined, we simply write

$$h : N \rightarrow \mathcal{P},$$
$$(I, \theta, \lambda) \mapsto (\gamma(I), I, s(\theta)) + \Sigma_{k=1}^{2n+l} \lambda_k d_k'(I, \theta), \qquad (2.78)$$

where $I_1, \ldots, I_m, \theta_1, \ldots, \theta_l, \lambda_1, \ldots, \lambda_{2n+l}$ are local coordinates in N as shown in (2.70).

3

Persistence of Overflowing Invariant Manifolds— Fenichel's Theorem

We are now at the point where we can state some general results on invariant manifolds of ordinary differential equations. As mentioned earlier, we will follow Fenichel's development of the theory since he explicitly treats the case of invariant manifolds with boundary which are often encountered in applications.

We consider a general autonomous ordinary differential equation defined on $\mathbb{R}^n$:

$$\dot{x} = f(x), \qquad x \in \mathbb{R}^n, \tag{3.1}$$

where f is a C^r, $r \geq 1$, function of x. Let us denote the flow generated by (3.1) by $\phi_t(p)$; i.e., $\phi_t(p)$ denotes the solution of (3.1) passing through the point $p \in \mathbb{R}^n$ at $t = 0$. We remark that $\phi_t(p)$ need not be defined for all $t \in \mathbb{R}$ or all $p \in \mathbb{R}^n$. Let $\bar{M} \equiv M \cup \partial M$ be a compact, connected C^r manifold with boundary contained in $\mathbb{R}^n$.

Definition 3.0.1 (a) $\bar{M} \equiv M \cup \partial M$ is said to be overflowing invariant under (3.1) if for every $p \in \bar{M}$, $\phi_t(p) \in \bar{M}$ for all $t \leq 0$ and the vector field (3.1) is pointing strictly outward on ∂M. (b) $\bar{M} \equiv M \cup \partial M$ is said to be inflowing invariant under (3.1) if for every $p \in \bar{M}$, $\phi_t(p) \in \bar{M}$ for all $t \geq 0$ and the vector field (3.1) is pointing strictly inward on ∂M. (c) $\bar{M} \equiv M \cup \partial M$ is said to be invariant under (3.1) if for every $p \in \bar{M}$, $\phi_t(p) \in \bar{M}$ for all $t \in \mathbb{R}$.

We make the following remarks concerning this definition.

1. The phrase "the vector field (3.1) is pointing strictly outward and is nonzero on ∂M" means that for every $p \in \partial M$, $\phi_t(p) \notin \bar{M}$ for all $t > 0$. A similar definition is obtained for the "... pointing strictly inward ..." by reversing time.

2. Overflowing invariant manifolds become inflowing invariant under time reversal and vice versa.

3. Since $\bar{M}$ is compact, $\phi_t(\cdot)\big|_{\bar{M}}$ exists for all $t \leq 0$ if $\bar{M}$ is overflowing invariant, for all $t \geq 0$ if $\bar{M}$ is inflowing invariant, and for all $t \in \mathbb{R}$ if $\bar{M}$ is invariant.

4. $\bar{M}$ can be an invariant manifold only if the vector field (3.1) is identically zero on ∂M, if $\partial M = \emptyset$, or if the vector field (3.1) is tangent to ∂M.

The following definition will also be useful.

Definition 3.0.2 *Let* $M \subset \mathbb{R}^n$ *be a compact, connected* C^r *manifold with boundary in* $\mathbb{R}^n$*. We say that* M *is locally invariant under (3.1) if for each* $p \in M$ *there exists a time interval* $I_p = \{t \in \mathbb{R} | t_1 < t < t_2$ *where* $t_1 < 0 < t_2\}$ *such that* $\phi_t(p) \in M$ *for all* $t \in I_p$.

We remark that the overflowing and inflowing invariant manifolds of Definition 3.0.1 are examples of locally invariant manifolds. The locally invariant manifold is *overflowing invariant* if we can take $t_2 = 0$ and $t_1 = -\infty$, and it is *inflowing invariant* if we can take $t_1 = 0$ and $t_2 = +\infty$.

Some Notation and Results from Analysis

We want to gather here some notation and results from analysis that will be used repeatedly throughout the rest of this chapter. For more background we refer the reader to Dieudonné [1960] and Abraham et al. [1988].

Norms

$L(E, F)$ will denote the space of linear maps from a complete normed linear vector space E to a complete normed linear vector space F (or *Banach space*). Let $A \in L(E, F)$; then the norm of A, denoted $\| A \|$, is defined as

$$\| A \| \equiv \sup \left\{ \frac{\| Ae \|}{\| e \|} \,\Big|\, e \in E, \, e \neq 0 \right\}.$$

It is a simple verification using this definition to show that the following are equivalent definitions for $\| A \|$:

$$\begin{aligned} \| A \| &= \inf \{ M > 0 \,|\, \| Ae \| \leq M \| e \| \,\forall e \in E \} \\ &= \sup \{ \| Ae \| \,|\, \| e \| \leq 1 \} = \sup \{ \| Ae \| \,|\, \| e \| = 1 \}. \end{aligned}$$

In particular, we have the ubiquitous estimate

$$\| Ae \| \leq \| A \| \| e \|.$$

We denote the space of p linear maps from $\underbrace{E \times \cdots \times E}_{p \text{ times}}$ to F by $L^p(E, F)$.

If $A \in L^p(E, F)$, then we denote the norm of A by

$$\| A \| = \sup \left\{ \frac{\| A(e_1, \ldots, e_p) \|}{\| e_1 \| \cdots \| e_p \|} \,\Big|\, e_1, \ldots, e_k \neq 0 \right\}.$$

It is a simple matter to verify from this definition that the following are equivalent definitions for this norm:

$$
\begin{aligned}
\| A \| &= \inf \{ M > 0 \mid \| A(e_1, \ldots, e_p) \| \leq M \| e_1 \| \cdots \| e_p \| \} \\
&= \sup \{ \| A(e_1, \ldots, e_p) \| \mid \| e_1 \| \leq 1, \ldots, \| e_p \| \leq 1 \} \\
&= \sup \{ \| A(e_1, \ldots, e_p) \| \mid \| e_1 \| = \cdots = \| e_p \| = 1 \} .
\end{aligned}
$$

If F is a complete normed linear vector space, then so is $L(E, F)$ and $L^p(E, F)$. Complete normed linear vector spaces are important to us because on them the contraction mapping theorem holds. We will also make use of the isomorphism $L^p(E, F) \simeq L(E, L^{p-1}(E, F))$.

Next we want to describe our notation for derivatives and norms of derivatives of C^r maps. We will not define the concept of the derivative of a map; this can be found in any analysis textbook (see, e.g., Dieudonné [1960] and Abraham et al. [1988]).

Let $U \subset E$ be an open set and consider a C^r map

$$
\begin{aligned}
f : U &\rightarrow F, \\
x &\mapsto Df(x).
\end{aligned}
$$

Then the derivative of f, denoted Df, is a map

$$
Df : U \rightarrow L(E, F);
$$

We define the norm of Df as follows

$$
\| Df \| \equiv \sup_{x \in U} \| Df(x) \| .
$$

Similarly, the pth derivative of f, denoted $D^p f$, $p \leq r$, is a map

$$
D^p f : U \rightarrow L^p(E, F);
$$

we define the norm of $D^p f$ as follows:

$$
\| D^p f \| \equiv \sup_{x \in U} \| D^p f(x) \| .
$$

The higher order derivatives are defined inductively with

$$
D^p f = D \left(D^{p-1} f \right) : U \rightarrow L \left(E, L^{p-1}(E, F) \right) .
$$

The Mean Value Inequality and Taylor's Theorem

For $x, x' \in U$, where we further assume that U is *convex*, we have

$$
\| f(x) - f(x') \| \leq \| Df \| \, \| x - x' \|
$$

where, as defined above,

$$\| Df \| \equiv \sup_{x \in U} \| Df(x) \| .$$

If $\| Df \|$ is uniformly bounded from above by some constant M, then

$$\| f(x) - f(x') \| \le M \| x - x' \| .$$

This inequality will prove particularly useful for obtaining upper bounds. On occasion we will also need to obtain lower bounds. For this, a Taylor expansion will usually be adequate:

$$f(x') = f(x) + Df(x)(x - x') + \mathcal{O}\left(\| x - x' \|^2\right)$$

where $\mathcal{O}(\cdot)$ denotes the standard "big oh" notation.

Obtaining a Lower Bound with a Taylor Expansion: We show how one can obtain a positive lower bound for $\| f(x) - f(x') \|$, for $\| x - x' \|$ sufficiently small, using a Taylor expansion. We begin with

$$f(x') - f(x) = Df(x)(x - x') + \mathcal{O}(\| x - x' \|^2).$$

If $Df(x)$ is invertible, then we have

$$(Df(x))^{-1} \left(f(x') - f(x)\right) = (x - x') + (Df(x))^{-1} \mathcal{O}(\| x - x' \|^2).$$

Taking the norm of this expression, we can then easily derive the following inequality:

$$\| (Df(x))^{-1} \| \| f(x') - f(x) \| \ge \| (Df(x))^{-1} (f(x') - f(x)) \|$$

$$\ge | \| x - x' \| - \| (Df(x))^{-1} \mathcal{O}(\| x - x' \|^2) \| |$$

or

$$\| f(x') - f(x) \| \ge \| (Df(x))^{-1} \|^{-1} | \| x - x' \| - \| (Df(x))^{-1} \mathcal{O}(\| x - x' \|^2) \| |.$$

We choose x and x' sufficiently close such that

$$\min \left\{ \| x - x' \| - \| (Df(x))^{-1} \mathcal{O}(\| x - x' \|^2) \| , \right.$$

$$\left. \| x - x' \| - \| \mathcal{O}(\| x - x' \|^2) \| \right\} > 0.$$

In this case we have

$$\| f(x') - f(x) \| \geq \| (Df(x))^{-1} \|^{-1} \| x - x' \|$$

$$- \| (Df(x))^{-1} \|^{-1} \| (Df(x))^{-1} \mathcal{O}(\| x - x' \|^2) \|$$

$$\geq \| (Df(x))^{-1} \|^{-1} \| x - x' \|$$

$$- \| (Df(x))^{-1} \|^{-1} \| (Df(x))^{-1} \| \| \mathcal{O}(\| x - x' \|^2) \|$$

$$= \| (Df(x))^{-1} \|^{-1} \| x - x' \| - \| \mathcal{O}(\| x - x' \|^2) \| \, .$$

Thus, we have obtained a *positive* lower bound for $\| f(x) - f(x') \|$ when $\| x - x' \|$ is small enough.

Topologies

$C^0(S, T)$ denotes the space of continuous maps between topological spaces S and T. When we speak of the C^0 topology we will be referring to the topology induced by the usual "sup" norm (when this norm can be appropriately defined).

We consider the main example for our work. Let $U \subset E$ be an open set and consider a C^r map

$$f : U \to F,$$

where E and F are complete normed linear vector spaces with topology induced by the norm. Then the C^0 norm of f is defined by

$$\sup_{x \in U} \| f \|_0 \, .$$

Inherited Structure from $\mathbb{R}^n$

We will be dealing with manifolds that are embedded in $\mathbb{R}^n$, which is equipped with the standard, and familiar, Euclidean inner product, norm, and metric. In all cases, our complete normed linear spaces E and F will be $\mathbb{R}^m$ and $\mathbb{R}^k$, for some m, $k \leq n$. As a result, the norms on $\mathrm{L}(E, F)$ and $\mathrm{L}^p(E, F)$ will also be constructed from the Euclidean norm by the recipes given above.

A Potential Notational Ambiguity: Following Fenichel's notation, $\dot{x} = f(x)$ denotes a vector field on $\mathbb{R}^n$; hence, x denotes a typical point in $\mathbb{R}^n$. Shortly, we will develop local coordinate charts for neighborhoods of regions of the overflowing invariant manifold M. This will result in a description of points near a portion of M

by ordered pairs (x, y) in Euclidean space where $(x, 0)$ represents a point on the manifold in the local coordinate chart. Obviously, in this context, x should not be confused with a general point in the ambient space of the manifold. This should be clear from the context. We have used this notation in order to stay as close as possible to Fenichel's original notation.

3.1   Overflowing Invariant Manifolds and Generalized Lyapunov-Type Numbers

Let $\bar{M} \equiv M \cup \partial M \subset \mathbb{R}^n$ be a compact, connected, C^r manifold with boundary and suppose that $\bar{M}$ is overflowing invariant. We want to characterize stability properties of $\bar{M}$ under the linearized dynamics in terms of the growth rates of vectors tangent to $\bar{M}$ and normal to $\bar{M}$. First we must address a technical detail arising from the fact that M has a boundary.

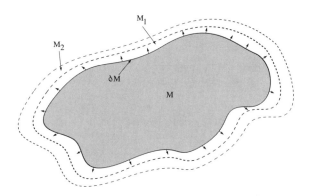

FIGURE 3.1. The Overflowing Invariant Manifolds, M, M_1, and M_2.

Technical Detail: In order to avoid technical details at the boundary we will allow M to "grow" a bit under the action of the flow $\phi_t(\cdot)$. By the compactness of $\bar{M}$, it follows that $\phi_t|_{\bar{M}}$ exists for all $t \leq 0$. By a change of time scale we can assume that $\phi_t|_{\bar{M}}$ exists for all $t \leq 3$ and we define

$$M_1 \equiv \phi_1(M), \qquad M_2 \equiv \phi_2(M).$$

It follows by existence and uniqueness of solutions of ordinary differential equations that $\bar{M}_1$ and $\bar{M}_2$ are overflowing invariant manifolds in $\mathbb{R}^n$; see Fig. 3.1 for a heuristic illustration.

With the standard (Euclidean) metric on $\mathbb{R}^n$, and the norm derived from this metric, we have the splitting

$$T\mathbb{R}^n|_{M_2} = TM_2 \oplus N,$$

where N is the normal bundle. We define the following orthogonal projection operator:

$$\Pi \ : \ T\mathbb{R}^n|_{M_2} \longrightarrow N.$$

Consider the following linear operators constructed from the linearized flow:

$$A_t(p) \equiv D\phi_{-t}|_{M_2}(p) \quad : \quad T_p M_2 \longrightarrow T_{\phi_{-t}(p)} M_2,$$
$$B_t(p) \equiv \Pi D\phi_t(\phi_{-t}(p))|_N \quad : \quad N_{\phi_{-t}(p)} \longrightarrow N_p.$$

For a point $p \in M_2$ we consider the following *nonzero* vectors:

$$\begin{aligned} w_0 &\in& N_p, \\ v_0 &\in& T_p M_2 \end{aligned}$$

and

$$\begin{aligned} w_{-t} &=& \Pi D\phi_{-t}(p)w_0, \\ v_{-t} &=& D\phi_{-t}(p)v_0. \end{aligned}$$

Notation: Generally, p will denote a typical point on $\bar{M}_2$. Vectors in $T_p\bar{M}_2$ will have a subscript 0. Accordingly, tangent vectors at the basepoint $\phi_{-t}(p)$ will have the subscript $-t$. We point this out since our notation for tangent vectors will not explicitly display the basepoint.

3.1.1 "Stability" of M and Generalized Lyapunov-Type Numbers

The notion of stability of M requires thoughtful consideration due to the boundary of M. Intuitively, an "object" is stable if trajectories through nearby points stay close to the "object" as $t \to \infty$. However, trajectories on M may leave M by passing through the boundary of M. Thus, trajectories starting close to M may get closer and closer to M; however, as they approach the boundary of M they may leave a neighborhood of M as we illustrate in Fig. 3.2.

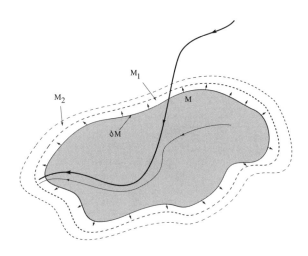

FIGURE 3.2. A Trajectory Near M.

We will say that M is stable under the linearized dynamics if the following limit holds for all $p \in M$, $w_0 \in N_p$:

$$\lim_{t \to \infty} \frac{\| w_0 \|}{\| w_{-t} \|} = 0.$$

Geometrically, this means that under the linearized dynamics vectors normal to M grow in backward time; see Fig. 3.3.

Smoothing properties of the flow near M will be characterized by the existence of the following limit for all $p \in M$, $w_0 \in N_p$, $v_0 \in T_p M$:

$$\lim_{t \to \infty} \left(\| w_0 \|^b / \| v_0 \| \right) / \left(\| w_{-t} \|^b / \| v_{-t} \| \right) = 0.$$

For $b = 1$, geometrically this means that neighborhoods of points on backward orbits "flatten out" under the linearized dynamics as they are carried forward in time. For $b < 1$, this flattening occurs at a faster rate. See Fig. 3.3 for a geometrical illustration.

For the purpose of calculation it will prove convenient to phrase these limits more quantitatively in terms of rates of convergence. We define the following *generalized Lyapunov-type numbers*:

$$\nu(p) = \inf \left\{ a : \left(\frac{\| w_0 \|}{\| w_{-t} \|} \right) / a^t \to 0 \quad \text{as} \quad t \uparrow \infty \quad \forall w_0 \in N_p \right\}. \quad (3.2)$$

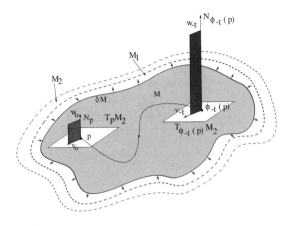

FIGURE 3.3. Behavior of Tangent Vectors under the Linearized Dynamics.

If $\nu(p) < 1$, we then define

$$\sigma(p) = \inf \left\{ b : \left(\| w_0 \|^b / \| v_0 \| \right) / \left(\| w_{-t} \|^b / \| v_{-t} \| \right) \to 0 \right.$$
$$\left. \text{as} \quad t \uparrow \infty \quad \forall v_0 \in T_p M_2, \quad w_0 \in N_p \right\}. \tag{3.3}$$

The following lemma will be useful.

Lemma 3.1.1

1. $\nu(p) = \overline{\lim}_{t \to \infty} \| B_t(p) \|^{\frac{1}{t}}$,

2. $\sigma(p) = \overline{\lim}_{t \to \infty} \dfrac{\log \| A_t(p) \|}{- \log \| B_t(p) \|}$, *if* $\nu(p) < 1$.

Proof: We begin with (1). Note that $w_0 = B_t(p) w_{-t}$, so that from (3.2) we have

$$\nu(p) = \inf \left\{ a : \left(\frac{\| B_t(p) w_{-t} \|}{\| w_{-t} \|} \right) / a^t \to 0 \quad \text{as} \quad t \uparrow \infty \quad \forall w_0 \in N_p \right\}. \tag{3.4}$$

With the norm of $B_t(p)$ defined as

$$\| B_t(p) \| \equiv \sup_{w_{-t} \in N_{\phi_{-t}(p)}} \frac{\| B_t(p) w_{-t} \|}{\| w_{-t} \|},$$

it immediately follows that

$$\nu(p) = \inf \left\{ a :\| B_t(p) \| / a^t \to 0 \quad \text{as} \quad t \uparrow \infty \right\}. \tag{3.5}$$

We will argue that from this it follows that

$$\nu(p) = \varlimsup_{t \to \infty} \| B_t(p) \|^{\frac{1}{t}} .$$

By definition we have

$$\varlimsup_{t \to \infty} \| B_t(p) \|^{\frac{1}{t}} = \inf_{T} \sup_{t \geq T} \| B_t(p) \|^{\frac{1}{t}} .$$

If a is such that $\lim_{t \to \infty}(\| B_t(p) \| / a^t) = 0$, then there exists T' such that $\| B_t(p) \|^{\frac{1}{t}} < a$ for all $t \geq T'$. Let $a(T) = \sup_{t \geq T} \| B_t(p) \|^{\frac{1}{t}}$. Thus, we have $a \geq a(T')$ for some T' from which it follows that $\inf a \geq \inf a(T)$. In order to show that $\nu(p) \equiv \inf a = \inf a(T)$ we argue by contradiction.

Assume that $\inf a > \inf a(T)$. We will show that there exists a constant c such that $\inf a > c > \inf a(T)$ such that $\lim_{t \to \infty}(\| B_t(p) \| / c^t) = 0$, thus contradicting expression (3.5) for $\nu(p)$. We begin by choosing c such that $\inf a > c > \inf a(T)$. Let $a(T')$ be such that $\inf a > c > a(T') \geq \inf a(T)$. But then

$$\lim_{t \to \infty} \frac{\| B_t(p) \|}{c^t} = \lim_{t \to \infty} \frac{\| B_t(p) \|}{(a(T'))^t} \left(\frac{a(T')}{c} \right)^t = 0$$

since

$$\frac{\| B_t(p) \|}{(a(T'))^t} \leq 1, \qquad \forall t > T',$$

and

$$\lim_{t \to \infty} \left(\frac{a(T')}{c} \right)^t = 0.$$

But this contradicts the initial assumption that $\nu(p) = \inf a$ is less than or equal to any a such that $\lim_{t \to \infty}(\| B_t(p) \| / a^t) = 0$.

A similar argument holds for (2). From (3.3) we can easily show that

$$\sigma(p) = \inf \left\{ b :\| A_t(p) \| \| B_t(p) \|^b \to 0 \quad \text{as} \quad t \uparrow \infty \right\}.$$

Consider an arbitrary b such that the limit $\| A_t(p) \| \| B_t(p) \|^b \to 0$ is attained; then there exists T such that for all $t > T$

$$0 < \| A_t(p) \| \| B_t(p) \|^b < 1.$$

Now for $\nu(p) < 1$ we have that for sufficiently large t $\log \| B_t(p) \| < 0$. Hence, taking natural logarithms of this expression gives

$$\log \frac{1}{\| A_t(p) \| \| B_t(p) \|^b} > 0,$$

from which it follows that

$$\frac{\log \| A_t(p) \|}{-\log \| B_t(p) \|} < b.$$

From this expression it follows that

$$\sigma(p) = \overline{\lim_{t \to \infty}} \frac{\log \| A_t(p) \|}{-\log \| B_t(p) \|},$$

and the argument proceeds as follows.

From the inequality

$$\frac{\log \| A_t(p) \|}{-\log \| B_t(p) \|} < b, \quad \forall t > T',$$

we can immediately conclude that

$$\sigma(p) = \inf \left\{ b \mid \| A_t(p) \| \| B_t(p) \|^b \to 0, \quad \text{as} \quad t \uparrow \infty \right\}$$

$$\geq \inf_T \sup_{t \geq T} \frac{\log \| A_t(p) \|}{-\log \| B_t(p) \|} \equiv \inf_T b(T)$$

$$\equiv \overline{\lim}_{t \to \infty} \frac{\log \| A_t(p) \|}{-\log \| B_t(p) \|}.$$

Now we argue by contradiction. Assume that

$$\sigma(p) = \inf b > \inf_T b(T).$$

Then for any constant c such that $\sigma(p) = \inf b > c > b(T') \geq \inf_T b(T)$ we have

$$\lim_{t \to \infty} \| A_t(p) \| \| B_t(p) \|^c = \lim_{t \to \infty} \| A_t(p) \| \| B_t(p) \|^{c \frac{b(T')}{b(T')}}$$

$$= \lim_{t \to \infty} \| A_t(p) \| \| B_t(p) \|^{b(T')} \| B_t(p) \|^{\frac{c}{b(T')}}.$$

Now for $t > T'$ we have

$$0 < \| A_t(p) \| \| B_t(p) \|^{b(T')} \leq 1,$$

and the condition $\nu(p) < 1$ implies that

$$\lim_{t \to \infty} \| B_t(p) \|^d = 0, \quad \forall d > 0.$$

From this we conclude that

$$\lim_{t \to \infty} \| A_t(p) \| \| B_t(p) \|^{b(T')} = 0,$$

which contradicts our original assumptions. Thus, the proof is complete.

□

Note that if $\nu(p) < 1$, for all $p \in M$, then M is "attracting" in the sense described above.

3.1.2 SOME PROPERTIES OF GENERALIZED LYAPUNOV-TYPE NUMBERS

In the following we prove some useful properties of generalized Lyapunov-type numbers.

Lemma 3.1.2 *Generalized Lyapunov-type numbers are constant on orbits, i.e.,*

$$\nu(p) = \nu(\phi_{-t}(p)), \qquad \sigma(p) = \sigma(\phi_{-t}(p)).$$

Proof: We first give the proof for $\nu(p)$. We recall the definition of $B_t(p)$:

$$B_t(p) = \Pi D\phi_t(\phi_{-t}(p))|_N : N_{\phi_{-t}(p)} \longrightarrow N_p.$$

From this we can easily deduce the following relations.

1. $B_{t+\tau}(p) = \Pi D\phi_{(t+\tau)}(\phi_{-(t+\tau)}(p))|_N : N_{\phi_{-(t+\tau)}(p)} \longrightarrow N_p,$

2. $B_\tau(p)B_t(\phi_{-\tau}(p))$
    $= \Pi D\phi_\tau(\phi_{-\tau}(p)) \, \Pi D\phi_t(\phi_{-(t+\tau)}(p)) : N_{\phi_{-(t+\tau)}(p)} \longrightarrow N_p.$

Using (1) and (2) and the fact that $T_{\phi_{-\tau}(p)} M_2$ is in the kernel of $\Pi D\phi_\tau(\phi_{-\tau}(p))$, it follows that

$$B_{t+\tau}(p) = B_\tau(p)B_t(\phi_{-\tau}(p)). \tag{3.6}$$

We fix $\tau > 0$, take the norm of both sides of this expression, and raise the result to the $1/t$ power to obtain

$$\| B_{t+\tau}(p) \|^{\frac{1}{t}} \leq \| B_\tau(p) \|^{\frac{1}{t}} \| B_t(\phi_{-\tau}(p)) \|^{\frac{1}{t}}. \tag{3.7}$$

Note that $B_\tau(p)$ is bounded, which implies that

$$\overline{\lim_{t \to \infty}} \| B_\tau(p) \|^{\frac{1}{t}} = 1. \tag{3.8}$$

We also have the following relations:

$$\overline{\lim_{t\to\infty}} \parallel B_{t+\tau}(p) \parallel^{\frac{1}{t}} = \overline{\lim_{t+\tau\to\infty}} \parallel B_{t+\tau}(p) \parallel^{\frac{1}{t+\tau}} \equiv \nu(p) \qquad (3.9)$$

and

$$\overline{\lim_{t\to\infty}} \parallel B_t\left(\phi_{-\tau}(p)\right) \parallel^{\frac{1}{t}} \equiv \nu\left(\phi_{-\tau}(p)\right). \qquad (3.10)$$

Then (3.7)–(3.10) imply that

$$\nu(p) \le \nu\left(\phi_{-\tau}(p)\right). \qquad (3.11)$$

Next observe that $B_\tau(p)$ is invertible; hence, (3.6) implies that

$$(B_\tau(p))^{-1} B_{t+\tau}(p) = B_t(\phi_{-\tau}(p)). \qquad (3.12)$$

Applying the same argument to (3.12), we obtain

$$\nu\left(\phi_{-\tau}(p)\right) \le \nu(p), \qquad (3.13)$$

which, together with (3.11), concludes the proof for $\nu(p)$.

The proof for $\sigma(p)$ follows similarly and we leave the details to the reader.
□

The significance of this result is that the generalized Lyapunov-type numbers depend only on the *backward limit sets* on M.

Proposition 3.1.3 $\nu(p)$ *is independent of the choice of metric on* $\mathbb{R}^n$ *and the splitting* $T\mathbb{R}^n|_{M_2} = TM_2 \oplus N$ *obtained with this metric. If* $\nu(p) < 1$, *then* $\sigma(p)$ *is also independent of the metric and splitting.*

Proof: We follow Fenichel [1971]. Suppose we have two metrics on $\mathbb{R}^n$ with associated norms, splittings, and generalized Lyapunov-type numbers. We denote these as follows:

$$
\begin{array}{cc}
(\cdot,\cdot) & (\cdot,\cdot)' \\
\parallel\cdot\parallel & \parallel\cdot\parallel' \\
T\mathbb{R}^n|_{M_2} = TM_2 \oplus N & T\mathbb{R}^n|_{M_2} = TM_2 \oplus N' \\
\nu, \sigma & \nu', \sigma'.
\end{array}
$$

Technical Detail: As a shorthand notation we use the notation

$$v - w,$$

where $v \in T\mathbb{R}^n|_{\bar{M}_2}$ and $w \in T\bar{M}_2$. We need to explain this notation, which literally means the subtraction of two vector bundle elements. This subtraction is really meant pointwise according to fibers; i.e., we only subtract elements if they have the same basepoint in the base space (which in our case will be $\bar{M}_2$, $\bar{M}_1$, or $\bar{M}$).

We must show that

$$\| B_t(p) \| / a^t \to 0 \quad \text{if and only if} \quad \| B'_t(p) \|' / a^t \to 0,$$

where $B'_t(p) \equiv \Pi' D\phi_t(\phi_{-t}(p))|_{N'}$, and

$$\| A_t(p) \| \, \| B_t(p) \|^b \to 0 \quad \text{if and only if} \quad \| A_t(p) \|' \, \| B'_t(p) \|'^b \to 0.$$

Any two norms on $\mathbb{R}^n$ are uniformly equivalent over the compact set $\bar{M}_2$ (see Hirsch and Smale [1974]); i.e., there exists a constant $c > 0$ such that

$$\frac{1}{c} \| v \| \leq \| v \|' \leq c \| v \| \tag{3.14}$$

for any $v \in T\mathbb{R}^n|_{\bar{M}_2}$. From (3.14) we have

$$\frac{1}{c} \| A_t(p)v \| \leq \| A_t(p)v \|' \leq c \| A_t(p)v \| . \tag{3.15}$$

Using (3.14) and (3.15) gives

$$\frac{1}{c^2} \frac{\| A_t(p)v \|}{\| v \|} \leq \frac{\| A_t(p)v \|'}{\| v \|'} \leq c^2 \frac{\| A_t(p)v \|}{\| v \|}. \tag{3.16}$$

Taking the supremum of this expression gives

$$\frac{1}{c^2} \| A_t(p) \| \leq \| A_t(p) \|' \leq c^2 \| A_t(p) \| . \tag{3.17}$$

Now for every $v \in T\mathbb{R}^n|_{\bar{M}_2}$ there exists $w \in T\bar{M}_2$ such that

$$\Pi v = v - w. \tag{3.18}$$

But

$$\| \Pi' v \|' \equiv \inf_{x \in T\bar{M}_2} \| v - x \|' \leq \| v - w \|'$$
$$\leq c \| v - w \| = c \| \Pi v \| . \tag{3.19}$$

Next we show the opposite inequality. For every $v \in T\mathbb{R}^n|_{\bar{M}_2}$ there exists $w \in T\bar{M}_2$ such that

$$\Pi' v = v - w. \tag{3.20}$$

Hence,

$$\| \Pi v \| \equiv \inf_{x \in T\bar{M}_2} \| v - x \| \leq \| v - w \|$$
$$\leq c \| v - w \|' = c \| \Pi' v \|' . \tag{3.21}$$

Using (3.19) and (3.21) we obtain

$$\frac{1}{c} \parallel \Pi v \parallel \leq \parallel \Pi' v \parallel' \leq c \parallel \Pi v \parallel . \tag{3.22}$$

In exactly the same way we derived (3.15) we obtain the following expression:

$$\frac{1}{c} \parallel B_t(p)v \parallel \leq \parallel B'_t(p)v \parallel' \leq c \parallel B_t(p)v \parallel . \tag{3.23}$$

Using (3.22) and (3.23) we obtain

$$\frac{1}{c^2} \parallel B_t(p) \parallel \leq \parallel B'_t(p) \parallel' \leq c^2 \parallel B_t(p) \parallel . \tag{3.24}$$

From (3.24) it immediately follows that

$$\parallel B_t(p) \parallel /a^t \to 0 \quad \text{if and only if} \quad \parallel B'_t(p) \parallel' /a^t \to 0.$$

Hence, $\nu(p) = \nu'(p)$. Similarly, using (3.17) and (3.24) we can show that

$$\parallel A_t(p) \parallel \parallel B_t(p) \parallel^b \to 0 \quad \text{if and only if} \quad \parallel A_t(p) \parallel' \parallel B'_t(p) \parallel'^b \to 0,$$

and from this expression conclude that $\sigma(p) = \sigma'(p)$. $\quad\square$

The following *uniformity lemma* of Fenichel allows us to make uniform estimates on the norms of the linear operators $A_t(\cdot)$ and $B_t(\cdot)$ based on the asymptotic (in time) information provided by the generalized Lyapunov-type numbers.

Lemma 3.1.4 (Uniformity Lemma, Fenichel, 1971)

1. Suppose

$$\parallel B_t(p) \parallel /a^t \to 0 \quad as \quad t \uparrow \infty \quad for\ every \quad p \in \bar{M}_1.$$

Then there are constants $\hat{a} < a$ and C such that

$$\| B_t(p) \| < C\hat{a}^t \quad \text{for every} \quad p \in \bar{M}_1 \quad \text{and} \quad t \geq 0.$$

2. *Under the hypotheses of (1), suppose also that $a \leq 1$ and*

$$\| A_t(p) \| \; \| B_t(p) \|^b \to 0 \quad \text{as} \quad t \uparrow \infty \quad \text{for every} \quad p \in \bar{M}_1.$$

Then there are constants $\hat{b} < b$ and C such that

$$\| A_t(p) \| \; \| B_t(p) \|^{\hat{b}} < C \quad \text{for every} \quad p \in \bar{M}_1 \quad \text{and} \quad t \geq 0.$$

3. *If $\nu(p) < a \leq 1$ and $\sigma(p) < b$ for every $p \in \bar{M}_1$, then*

$$\| B_t(p) \| \to 0 \quad \text{as} \quad t \uparrow \infty$$

and

$$\| A_t(p) \| \; \| B_t(p) \|^b \to 0 \quad \text{as} \quad t \uparrow \infty$$

uniformly for every $p \in \bar{M}_1$.

4. *$\nu(\cdot)$ and $\sigma(\cdot)$ attain their suprema on M.*

Proof: We begin by proving (1) and (2) simultaneously. The reader can easily verify that the proof of (1) does not depend on the hypotheses of (2). Let a and b be given as in the hypotheses of (1) and (2). Then for each $p \in \bar{M}_1$ there is a number $T(p)$ such that

$$\| B_{T(p)}(p) \| < a^{T(p)},$$
$$\| A_{T(p)}(p) \| \; \| B_{T(p)}(p) \|^b < 1.$$

By continuity with respect to initial conditions, for every $p \in \bar{M}_1$ there is a neighborhood $U(p)$ of p in M_2 such that for all $p' \in \overline{U(p)}$,

$$\| B_{T(p)}(p') \| < a^{T(p)},$$
$$\| A_{T(p)}(p') \| \; \| B_{T(p)}(p') \|^b < 1.$$

$\bar{M}_1$ is compact so we can choose finitely many points $p_1, \ldots, p_N$ such that

$$\bar{M}_1 \subset U(p_1) \cup \cdots \cup U(p_N).$$

We choose $\hat{a} < a$ and $\hat{b} < b$ such that for $p' \in \overline{U(p_i)}$ we have

$$\| B_{T(p_i)}(p') \| < \hat{a}^{T(p_i)},$$

$$\| A_{T(p_i)}(p') \| \, \| B_{T(p_i)}(p') \|^{\hat{b}} < 1.$$

Let $p \in \bar{M}_1$ be given. Choose a (*nonunique*) sequence of integers as follows:

$$
\begin{aligned}
i_1 &: \quad p \in U(p_{i_1}), \\
i_2 &: \quad \phi_{-T(p_{i_1})}(p) \in U(p_{i_2}), \\
i_3 &: \quad \phi_{-(T(p_{i_1})+T(p_{i_2}))}(p) \in U(p_{i_3}), \\
&\quad \vdots \\
i_j &: \quad \phi_{-(T(p_{i_1})+\cdots+T(p_{i_{j-1}}))}(p) \in U(p_{i_j}).
\end{aligned}
$$

(Note: It should be clear that the elements of the set $\{p_{i_1}, \ldots, p_{i_j}\}$ are chosen from the elements of the set $\{p_1, \ldots, p_N\}$.) Let $\tau(j) = T(p_{i_1}) + \cdots + T(p_{i_j})$ and let $t > 0$ be given. Then we write $t > 0$ as

$$t = \tau(j) + r,$$

for some $j \geq 0$ (with $\tau(0) \equiv 0$) where $0 \leq r < \max T(p_i)$.

Now $B_t(p) = B_{T(p_{i_1})+\cdots+T(p_{i_j})+r}(p)$. Expanding this out gives the product

$$B_t(p) = B_{T(p_{i_1})}(p) \cdot B_{T(p_{i_2})}(\phi_{-T(p_{i_1})}(p))$$

$$\cdot B_{T(p_{i_3})}(\phi_{-(T(p_{i_1})+T(p_{i_2}))}(p)) \cdots B_{T(p_{i_j})}(\phi_{-(T(p_{i_1})+\cdots+T(p_{i_{j-1}}))}(p))$$

$$\cdot B_r(\phi_{-(T(p_{i_1})+\cdots+T(p_{i_j}))}(p)).$$

From this expression we easily obtain the bound

$$\| B_t(p) \| \; < \; \| B_r(\phi_{-(T(p_{i_1})+\cdots+T(p_{i_j}))}(p)) \| \left(\frac{\hat{a}^r}{\hat{a}^r}\right) \hat{a}^{T(p_{i_1})} \ldots \hat{a}^{T(p_{i_j})}$$

$$\leq \; C \hat{a}^t,$$

where

$$C \equiv \sup_{\substack{p \in \bar{M}_1 \\ 0 \leq r \leq \max T(p_i)}} \| B_r(p) \| / \hat{a}^r.$$

In exactly the same way, we estimate $\| A_t(p) \| \, \| B_t(p) \|^{\hat{b}}$. Given $p \in M_1$ and $t > 0$, we write $t = \tau(j) + r$ and

$$\| A_t(p) \| \| B_t(p) \|^{\hat{b}} \leq \| A_{T(p_{i_1})}(p) \| \| B_{T(p_{i_1})}(p) \|^{\hat{b}}$$

$$\cdot \| A_{T(p_{i_2})}(\phi_{-T(p_{i_1})}(p)) \| \| B_{T(p_{i_2})}(\phi_{-T(p_{i_1})}(p)) \|^{\hat{b}}$$

$$\cdots \| A_r(\phi_{-(T(p_{i_1})+\cdots+T(p_{i_j}))}(p)) \| \| B_r(\phi_{-(T(p_{i_1})+\cdots+T(p_{i_j}))}(p)) \|^{\hat{b}}$$

$$< \| A_r(\phi_{-(T(p_{i_1})+\cdots+T(p_{i_j}))}(p)) \| \| B_r(\phi_{-(T(p_{i_1})+\cdots+T(p_{i_j}))}(p)) \|^{\hat{b}} .$$

Letting

$$C \equiv \sup_{\substack{p \in M_1 \\ 0 \leq r \leq \max T(p_i)}} \| A_r(p) \| \| B_r(p) \|^{\hat{b}}$$

gives the result.

Part (3) of the uniformity lemma is an obvious consequence of parts (1) and (2). We now prove part (4) for ν. The result for σ follows the same argument, with obvious modifications.

The proof is by contradiction. Suppose ν does not attain its supremum on M. Then we denote

$$a = \sup_{p \in \bar{M}_1} \nu(p).$$

By part (1) of the uniformity lemma there exists $\hat{a} < a$ such that $\| B_t(p) \| < C\hat{a}^t$. Then for any $\hat{a} < a' < a$ we have

$$\frac{\| B_t(p) \|}{a'^t} \to 0 \quad \text{as} \quad t \uparrow \infty.$$

This implies that $\nu(p) < a'$, which is a contradiction.   □

Example 3.1.1.

We now give a simple and familiar example where we can explicitly calculate the generalized Lyapunov-type numbers. Consider the following linear vector field on the plane:

$$\begin{pmatrix} \dot{x} \\ \dot{y} \end{pmatrix} = \begin{pmatrix} \lambda & 0 \\ 0 & -\mu \end{pmatrix} \begin{pmatrix} x \\ y \end{pmatrix}, \qquad \lambda, \mu > 0. \tag{3.25}$$

It is a trivial calculation to verify that the following set is an overflowing invariant manifold:

$$M = \{(x, y) \mid y = 0, \ -1 < x < 1\}.$$

We have the splitting

$$T\mathbb{R}^2|_M = TM \oplus N,$$

where

$$
\begin{aligned}
TM &= \left\{ (x,0) \times (\mathbb{R}^1,0) \mid -1 < x < 1 \right\}, \\
N &= \left\{ (x,0) \times (0,\mathbb{R}^1) \mid -1 < x < 1 \right\},
\end{aligned}
$$

from which it easily follows that

$$\Pi = \begin{pmatrix} 0 & 0 \\ 0 & 1 \end{pmatrix}. \tag{3.26}$$

The matrix associated with the linearized flow is easily computed and found to be

$$D\phi_t = \begin{pmatrix} e^{\lambda t} & 0 \\ 0 & e^{-\mu t} \end{pmatrix}. \tag{3.27}$$

The linearized operators associated with the tangential and normal linearized dynamics are given by

$$
\begin{aligned}
A_t(p) &= D\phi_{-t}|_M(p) = \begin{pmatrix} e^{-\lambda t} & 0 \\ 0 & 0 \end{pmatrix}, \\
B_t(p) &= \Pi D\phi_t(\phi_{-t}(p)) = \begin{pmatrix} 0 & 0 \\ 0 & e^{-\mu t} \end{pmatrix}.
\end{aligned}
$$

The generalized Lyapunov-type numbers are easily computed and are given by

$$
\begin{aligned}
\nu(p) &= \overline{\lim_{t \to \infty}} \, |e^{-\mu t}|^{\frac{1}{t}} = e^{-\mu} < 1, \\
\sigma(p) &= \overline{\lim_{t \to \infty}} \, \frac{\log |e^{-\lambda t}|}{-\log |e^{-\mu t}|}, \\
&= \overline{\lim_{t \to \infty}} \, \frac{-\lambda t \log e}{\mu t \log e} = -\frac{\lambda}{\mu}.
\end{aligned}
$$

Since the vector field is linear, the generalized Lyapunov-type numbers are identical at each point.

End of Example 3.1.1

3.2   Coordinates and Dynamics Near $\bar{M}$

In order to do the necessary analysis to prove the invariant manifold theorems we will need to define coordinates in a neighborhood of $\bar{M}$ and
describe the dynamics in terms of these coordinates.

3.2.1   LOCAL COORDINATES NEAR $\bar{M}$

As discussed in Section 2.2.2, vector bundles have the property of *local
triviality*. The relevant implication of this property for our case is that
there exist open sets $U_i \in M_1$ such that for each $p \in U_i$ we can choose
an orthonormal basis for N_p and the basis will vary in a C^{r-1} fashion for
all $p \in U_i$. In this case we say that $N|_{U_i}$ has a C^{r-1} orthonormal basis.
The existence of a C^{r-1} orthonormal basis will be important for defining
coordinates near M_1. For this purpose we want to argue that the open
sets, U_i, in which N is locally trivial can also be chosen as the domains for
coordinates on M_1. We begin with a definition.

Definition 3.2.1 *Suppose* $\mathcal{U} = \{U_i : i \in \mathcal{I}\}$ *and* $\mathcal{V} = \{V_j : j \in \mathcal{J}\}$ *are two
open covers of* $\bar{M}_2$, *where* $\mathcal{I}$ *and* $\mathcal{J}$ *are index sets. We say that* $\mathcal{U}$ *is subordinate to* $\mathcal{V}$ *if, for every* $i \in \mathcal{I}$, *there exists a* $j \in \mathcal{J}$ *such that* $U_i \subset V_j$.

Now suppose $\mathcal{U} = \{U_i : i \in \mathcal{I}\}$ is an open cover of $\bar{M}_1$ such that, in each
U_i, $N|_{U_i}$ has a C^{r-1} orthonormal basis. The following proposition gives a
construction of coordinate charts in which N is locally trivial in each chart
domain (recall the definition of an *atlas* given after Definition 2.1.3).

Proposition 3.2.1 *For every open cover* $\mathcal{U}$ *of* $\bar{M}_2$ *there exists atlases*

$$\left\{ \left(U_i^j, \sigma_i \right) \; : \; i = 1, \dots, s \, ; \, j = 1, \dots, 6 \right\} \tag{3.28}$$

such that

$$U_i^1 \subset \overline{U}_i^1 \subset U_i^2 \subset \overline{U}_i^2 \subset U_i^3 \subset \overline{U}_i^3 \subset U_i^4 \subset \overline{U}_i^4 \subset U_i^5 \subset \overline{U}_i^5 \subset U_i^6 \subset \overline{U}_i^6 \tag{3.29}$$

with

$$\sigma_i \left(U_i^j \right) = \mathcal{D}^j, \qquad j = 1, \dots, 6 \tag{3.30}$$

where $\mathcal{D}^j \equiv \left\{ x \in \mathbb{R}^{n-k} \; : \| \, x \, \| < j \right\}$, *i.e., the open disc of radius* j. *More-
over, the open covers*

$$\mathcal{U}^j = \left\{ U_i^j \; : i = 1, \dots, s \right\} \tag{3.31}$$

are subordinate to $\mathcal{U}$.

Proof: The proof of this result consists of two steps.

Step 1. Show that at each $p \in \bar{M}_2$ there are coordinate charts (U_p^j, σ_p), $j = 1, \ldots, 6$ such that $\sigma_p(U_p^j) = \mathcal{D}^j$ with the U_p^j satisfying the nesting property of (3.29) and U_p^6 is contained in an element of $\mathcal{U}$. Then it will immediately follow that U_p^j, $j = 1, \ldots, 5$, are contained in the same element of $\mathcal{U}$.

Step 2. Show that the U_p^1 obtained in this way can be chosen so as to form a cover of $\bar{M}_2$. From the nesting property (3.29) it then follows that the U_p^j, $j = 1, \ldots, 5$, can also be chosen to form a cover of $\bar{M}_2$.

We begin with step 1. Let $p \in \bar{M}_2$ and let W be an element of the cover $\mathcal{U}$ containing p. Then there is a coordinate chart at p, (V_p, σ_p) and V_p can be chosen such that $V_p \subset W$. Without loss of generality we can assume $\mathcal{D}^6 \subset \sigma_p(V_p)$; if this is not the case, then we need only multiply σ_p by an appropriate scalar to make it so. Letting $U_p^6 \equiv \sigma_p^{-1}(\mathcal{D}^6)$, it is clear that (σ_p, U_p^6) is a coordinate chart around p.

Next we consider the sets

$$\sigma_p^{-1}(\mathcal{D}^j) \equiv U_p^j, \quad j = 1, \ldots, 5.$$

Since σ_p is a C^r diffemorphism, it follows that $U_p^j \subset U_p^k$ for $j \leq k$, because $\mathcal{D}^j \subset \mathcal{D}^k$, and trivially we have $U_p^j \subset \bar{U}_p^j$. Hence, we have shown that

$$U_p^1 \subset \bar{U}_p^1 \subset U_p^2 \subset \bar{U}_p^2 \subset U_p^3 \subset \bar{U}_p^3 \subset U_p^4 \subset \bar{U}_p^4 \subset U_p^5 \subset \bar{U}_p^5 \subset U_p^6 \subset \bar{U}_p^6 \tag{3.32}$$

with

$$\sigma_p\left(U_p^j\right) = \mathcal{D}^j, \quad j = 1, \ldots, 6. \tag{3.33}$$

Moreover, by construction, $U_p^j \subset W \in \mathcal{U}$ for $j = 1, \ldots, 6$.

Using the compactness of $\bar{M}_2$, we can extract a finite covering from this infinite covering, U_i^1, $i = 1, \ldots, s$. By construction, U_i^j, $i = 1, \ldots, s$, is also a covering of $\bar{M}_2$, for $j = 2, \ldots, 5$. Thus, $\{(U_i^j, \sigma_i) : i = 1, \ldots, s\}$ forms an atlas of $\bar{M}_2$, for each $j = 1, \ldots, 6$. Moreover, by construction, each U_i^j is contained in an element of $\mathcal{U}$. □

This result has an immediate corollary that arises directly from the construction.

Corollary 3.2.2 $N|_{U_i^j}$ *has a C^{r-1} orthonormal basis, for all i and j.*

Proof: This is an immediate consequence of the fact that by construction the open covers $\bigcup_{i=1}^s U_i^j$, for $j = 1, \ldots, 6$, are subordinate to the original cover $\mathcal{U}$ and the original cover can be chosen such that N, restricted to a set in the cover, is locally trivial. □

3.2.2   "Jiggling" the Normal Bundle to Gain a Derivative, and the Construction of a Neighborhood of M_1

An obvious choice for local coordinates near $\bar{M}_1$ would be to use local coordinates in the tangent bundle and normal bundle splitting (cf. Proposition 2.2.5). However, there is a problem with this as the normal bundle is only C^{r-1} and we will be seeking invariant manifolds that are C^r. However, a result of Whitney allows us to choose a C^r bundle transverse to TM_1 that can be used to provide appropriate coordinates. Let $k \equiv n - \dim M$. Then k is the dimension of the fibers of N and we have the following result.

Proposition 3.2.3 *There is a C^r k-dimensional bundle $N' \subset T\mathbb{R}^n|_{M_1}$, transversal to TM_1. Moreover, given $\epsilon > 0$, for any set U_i^j as constructed in Proposition 3.2.1, there exist orthonormal bases*

$$\left\{ e_1^{ij}(p), \dots, e_k^{ij}(p) \right\} \quad for \quad N|_{U_i^j},$$

$$\left\{ f_1^{ij}(p), \dots, f_k^{ij}(p) \right\} \quad for \quad N'|_{U_i^j},$$

such that

$$\| e_\ell^{ij}(p) - f_\ell^{ij}(p) \| < \epsilon, \quad \ell = 1, \dots, k.$$

The $f_\ell^{ij}(p)$ can be chosen to be C^r functions of $p \in U_i^j$.

Proof: This result can be found in Whitney [1936], Lemma 23, p. 667. We outline the proof as given in Fenichel [1970].

Let $G(k,n)$ denote the analytic manifold of all k planes through the origin in $\mathbb{R}^n$. We can view N as determined by a map $\Phi : \bar{M}_2 \to G(k,n)$, which assigns to each point $p \in \bar{M}_2$ the k plane through the origin parallel to N_p. The proposition will be proved if we show that there is a C^r map $\tilde{\Phi}$ arbitrarily C^0 close to Φ on $\bar{M}_1$.

> **Technical Note:** A *retraction* is a continuous map of a topological space X onto a subspace A which is the identity map on A. A subspace of a topological space X is a subset of X equipped with the relative topology. See Fuks and Rokhlin [1984] for a discussion of the existence of retractions.

Let i denote any C^∞ embedding of $G(k,n)$ into $\mathbb{R}^p$ for some appropriate $p > n$. Let σ be a C^∞ retraction of a neighborhood of $i\left(G(k,n)\right)$ in $\mathbb{R}^p$ into $i\left(G(k,n)\right)$. Then $i \circ \Phi$ maps $\bar{M}_2$ into $\mathbb{R}^p$. Next, extend $i \circ \Phi$ to a continuous function $\Phi_1 : \mathbb{R}^n \to \mathbb{R}^p$ by, say, the Tietze–Urysohn Extension Theorem. Then we use standard results from approximation theory (see, e.g., Llavona [1986] or Fuks and Rokhlin [1984]) to approximate Φ_1 uniformly by a C^r

function $\Phi_2 : {\rm I\!R}^n \to {\rm I\!R}^p$. Let $\tilde{\Phi} \equiv i^{-1} \circ \sigma \circ \Phi_2|_{\bar{M}_1}$. Then $\tilde{\Phi}$ is C^r and C^0 close to Φ on $\bar{M}_1$.   □

We will refer to N' as the *transversal bundle*. Elements of N' are denoted by the pairs (p, v), where $p \in M_1 \subset {\rm I\!R}^n$ and v is a tangent vector to ${\rm I\!R}^n$ at p transversal to $T_p M_1$.

Next we construct a neighborhood of the manifold M_1. For this purpose we define the map

$$h \ : \ N' \longrightarrow {\rm I\!R}^n,$$
$$(p, v) \longrightarrow p + v. \tag{3.34}$$

The following proposition is also due to Whitney.

Proposition 3.2.4 *Let $K \subset M_1$ be any compact subset. Then h is a C^r diffeomorphism from a neighborhood of the zero section of $N'|_K$ to a neighborhood of K in ${\rm I\!R}^n$.*

Proof: The proof goes exactly the same as Proposition 2.2.5 applied to the transversal bundle N'. See Whitney [1936], Lemma 23, p. 667.   □

We remark that it follows from Proposition 3.1.3 that the generalized Lyapunov-type numbers computed with respect to the splitting $T{\rm I\!R}^n|_{M_1} = TM_1 \oplus N$ are identical to those computed with respect to the splitting $T{\rm I\!R}^n|_{M_1} = TM_1 \oplus N'$.

We next want to define a neighborhood of $\bar{M}$ in such a way that we can control its "thickness" around $\bar{M}$. This will be important for making certain estimates. Let

$$N'_\epsilon = \{(p, v) \in N' : \| v \| < \epsilon\}.$$

Using Proposition 3.2.4, one can easily show that there is an $\epsilon_0 > 0$ such that for all $0 < \epsilon \le \epsilon_0$ the map h defined in (3.34) is a C^r map, mapping $N'_\epsilon|_{\cup_{i=1}^s U_i^5}$ onto a neighborhood of $\cup_{i=1}^s \bar{U}_i^4$. Generally, we will not distinguish between this neighborhood and $N'_\epsilon|_{\cup_{i=1}^s U_i^5}$.

For each i, we choose an orthonormal basis of $N'_\epsilon|_{U_i^6}$. and we define the map

$$\tau_i \ : \ N'_\epsilon|_{U_i^6} \longrightarrow {\rm I\!R}^k,$$
$$(p, v) \longrightarrow \left\{ \begin{array}{l} \text{vector of coordinates of } v \text{ with respect} \\ \text{to the chosen basis of } N'_\epsilon|_p \end{array} \right\}. \tag{3.35}$$

Since we have chosen an orthonormal basis, we have $\| \tau_i(p, v) \| = \| v \|$.

Bounds on the Coordinate Maps: For all i and $(p, v) \in N'_\epsilon|_{\bar{M}_2}$, we have the bounds

$$\| D\sigma_i(p) \| < c, \qquad \| D\sigma_i^{-1}(\sigma_i(p)) \| < c.$$

Derivatives along the fibers of $N'_\epsilon|_{\bar{M}_2}$ are simple since we are using an orthonormal basis to define $\tau_i(p, v)$. In this case we can take

$$\| D\tau_i(p, v) \| = \| D\tau_i^{-1}(\sigma_i(p), \tau_i(p, v)) \| = 1.$$

Next we define the map

$$\sigma_i \times \tau_i : N'_\epsilon|_{U_i^6} \longrightarrow \mathbb{R}^{n-k} \times \mathbb{R}^k$$

by

$$(\sigma_i \times \tau_i)(p, v) = (\sigma_i(p), \tau_i(p, v)) \equiv (x, y).$$

Proposition 3.2.5 *The map $\sigma_i \times \tau_i$ is a C^r diffeomorphism.*

Proof: A simple calculation of the local coordinate representative of the Jacobian of $\sigma_i \times \tau_i$ shows that it is nonsingular for all $(p, v) \in N'_\epsilon$. Hence, by the inverse function theorem, $\sigma_i \times \tau_i$ is a C^r diffeomorphism. □

Note that we have

$$(\sigma_i \times \tau_i)^{-1}(x, y) = (p, v)$$

and

$$(\sigma_i \times \tau_i)^{-1}(x, 0) = (p, 0).$$

Now if ϵ is small, $N'_\epsilon|_{U_i^5}$ is a C^r neighborhood of $\bar{U}_i^4$ in $\mathbb{R}^n$. Therefore, $\sigma_i \times \tau_i$ specifies a local coordinate system in $\mathbb{R}^n$ near part of M. Points in M are characterized by vanishing of the second component of this map. See Fig. 3.4 for a geometrical illustration.

We will need to express flows and their derivatives in terms of these local coordinates, and we now turn to this question.

3.2.3   Local Expressions for the Time-T Map Generated by the Flow

Henceforth, we will make the following assumption:

$$\boxed{\nu(p) < 1 \text{ and } \sigma(p) < \tfrac{1}{r} \text{ for all } p \in \bar{M}_1.}$$

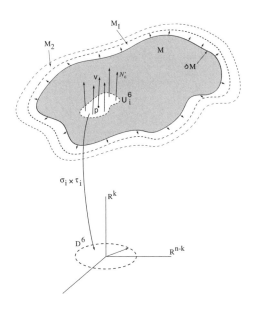

FIGURE 3.4. Local Coordinates Near M_2.

This will be in the hypotheses in our main theorem. It follows then by the uniformity lemma that we can find a T sufficiently large such that

$$\| B_T'(p) \| < \tfrac{1}{4}, \tag{3.36}$$

$$c^{2r} \| A_T(p) \|^r \| B_T'(p) \| < \tfrac{1}{4}, \tag{3.37}$$

for all $p \in \bigcup_{i=1}^{s} \overline{U}_i^5$. (Recall, $B_T'(p) \equiv \Pi' D\phi_T(\phi_{-T}(p))$.)

> **Technical Point:** The constant c^{2r} is the same c that bounds $\| D\sigma_i(p) \| < c$ and $\| D\sigma_i^{-1}(\sigma_i(p)) \| < c$ as described earlier. We introduce this at this point in order to make the local coordinate expressions for these estimates simpler later on.

We choose $T > 1$ such that these inequalities hold *and we keep this T fixed for the remainder of the discussion.* Note that if $0 \le k \le r$, we also have

$$c^{2k} \| A_T(p) \|^k \| B_T'(p) \| < \tfrac{1}{4}. \tag{3.38}$$

which follows from (3.36), (3.37), and the fact that $\| A_t(p) \|^k$ is a monotonically increasing function of k.

The condition $\| B'_T(p) \| < \frac{1}{4}$ means that $D\phi_T(\cdot)$ decreases the lengths of vectors in N' by a factor of 4. We next want to show that this contractive property holds for the nonlinear flow, with ϵ sufficiently small.

Lemma 3.2.6 *For ϵ sufficiently small we have*

$$\phi_T : N'_\epsilon|_{\phi_{-T}(\overline{U}_i^4)} \longrightarrow N'_{\epsilon/3}|_{U_i^5}. \tag{3.39}$$

Proof: The condition $\| B'_T(p) \| < \frac{1}{4}$ is related to the linearized dynamics transverse to M_1 and this lemma is a statement describing the nonlinear dynamics. Thus, we want to estimate the difference between the linear and nonlinear dynamics near M_1 for the finite time T. We let

$$x(t) = \phi_t(p) + \xi(t),$$

where $p \in \phi_{-T}(\overline{U}_i^4)$. We will think of $\phi_t(p)$ as representing motion on M_1 and ξ as representing a deviation from this motion, transverse to M_1. The linearized dynamics of ξ satisfies

$$\dot{\xi}_{\text{linear}} = Df(\phi_t(p))\xi_{\text{linear}}, \tag{3.40}$$

and the nonlinear dynamics of ξ is given by

$$\dot{\xi}_{\text{nonlinear}} = Df(\phi_t(p))\xi_{\text{nonlinear}} + \mathcal{O}\left(\| \xi_{\text{nonlinear}} \|^2\right). \tag{3.41}$$

Thus, from a simple Gronwall estimate we obtain

$$|\xi_{\text{linear}}(T) - \xi_{\text{nonlinear}}(T)| = \mathcal{O}\left(\| \xi_{\text{nonlinear}} \|^2\right).$$

Hence, for $\xi = \mathcal{O}(\epsilon)$, we see that the difference in the nonlinear and linear dynamics near M_1 for finite time T is $\mathcal{O}(\epsilon^2)$. Now for the linear dynamics we have

$$\Pi' D\phi_T\left(\phi_{-T}(\cdot)\right) : N'_\epsilon|_{\phi_{-T}(\overline{U}_i^4)} \longrightarrow N'_{\epsilon/4}|_{U_i^5};$$

so with this error estimate, it follows for the nonlinear dynamics that for ϵ sufficiently small

$$\phi_T : N'_\epsilon|_{\phi_{-T}(\overline{U}_i^4)} \longrightarrow N'_{\epsilon/3}|_{U_i^5}$$

since $\frac{\epsilon}{4} + \mathcal{O}(\epsilon^2) < \frac{\epsilon}{3}$. Thus, the lemma is proved

□

Local expressions for the time T map in the coordinates developed above are given by

$$(x, y) \mapsto \left(f_{ij}^0(x, y), g_{ij}^0(x, y)\right) \in \mathbb{R}^{n-k} \times \mathbb{R}^k,$$

where

$$f_{ij}^0(x,y) \equiv \sigma_j \circ \phi_T \circ (\sigma_i \times \tau_i)^{-1}(x,y), \tag{3.42}$$

$$g_{ij}^0(x,y) \equiv \tau_j \circ \phi_T \circ (\sigma_i \times \tau_i)^{-1}(x,y). \tag{3.43}$$

These expressions are defined for

$$(x,y) \in (\sigma_i \times \tau_i) \, N'_\epsilon|_{\overline{U}_i^4 \cap \phi_{-T}(\overline{U}_j^4)}.$$

See Fig. 3.5 for an illustration of the geometry.

> **Technical Note:** In Proposition 3.2.4 we constructed a neighborhood of the manifold and later (immediately following Corollary 3.2.2) made the remark "... we will not distinguish between this neighborhood and $N'_\epsilon|_{\cup_{i=1}^s U_i^5}$". We now want to explain one of the consequences of this remark.
>
> Strictly speaking, expressions (3.42) and (3.43) are not correct and should be written
>
> $$f_{ij}^0(x,y) \equiv \sigma_j \circ h^{-1} \circ \phi_T \circ h \circ (\sigma_i \times \tau_i)^{-1}(x,y),$$
> $$g_{ij}^0(x,y) \equiv \tau_j \circ h^{-1} \circ \phi_T \circ h \circ (\sigma_i \times \tau_i)^{-1}(x,y).$$
>
> Thus, not distinguishing the neighborhood of the manifold and $N'_\epsilon|_{\cup_{i=1}^s U_i^5}$ means that we ignore the h and h^{-1} terms in the local coordinate expressions for the time-T map. This presents no difficulties in the analysis (since our estimates will always be dealt with in the local coordinate form), so following Fenichel's original work, we will continue with this notation.

> **Technical Note:** Equation (3.42) requires a more careful explanation. Originally, σ_j was defined as a diffeomorphism with domain $U_j^6 \subset M_1$. However, the point $\phi_T (\sigma_i \times \tau_i)^{-1}(x,y)$ may not generally be in U_j^6, but rather it will lie in a neighborhood of U_j^6 in $\mathbb{R}^n$. Hence, one would wonder about the meaning of the expression $f_{ij}^0(x,y) \equiv \sigma_j \circ \phi_T \circ (\sigma_i \times \tau_i)^{-1}(x,y)$. To understand this, one should go back to the definition of the local coordinates. From this we see that it is implied in this expression that σ_j acts on the basepoint of $\phi_T \circ (\sigma_i \times \tau_i)^{-1}(x,y)$.

It is easy to verify that overflowing invariance of $\bar{M}$ implies that $g_{ij}^0(x,0) = 0$. Thus, $\left(f_{ij}^0(x,0),0\right)$ is the local expression for the time-T map *restricted to* M_1. We denote partial differentiation with respect to x and y by D_1 and D_2, respectively. Then local representations of A_T at the point $\phi_T(\sigma_i^{-1}(x))$ are given by $\left(D_1 f_{ij}^0(x,0)\right)^{-1}$ and local representations of

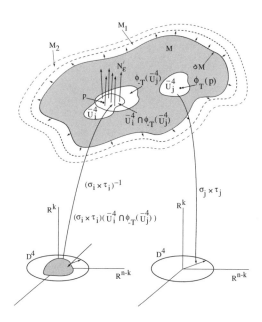

FIGURE 3.5. The Geometry Associated with Local Coordinate Expressions for the Time-T Map.

B_T at $\phi_T(\sigma_i^{-1}(x))$ are given by $D_2 g_{ij}^0(x,0)$. Rewriting (3.37) in these local expressions gives

$$\| \left(D_1 f_{ij}^0(x,0)\right)^{-1} \|^k \| D_2 g_{ij}^0(x,0) \| < \tfrac{1}{4} \qquad (3.44)$$

for all

$$(x,0) \in (\sigma_i \times \tau_i) \, N'_\epsilon |_{\overline{U}_i^4 \cap \phi_{-T}(\overline{U}_j^4)}$$

and $0 \le k \le r$. The local expression for (3.39) is manifested as

$$\| g_{ij}^0(x,y) \| < \frac{\epsilon}{3}. \qquad (3.45)$$

Note that there are no factors of c^{2k} (3.44). This is because introducing them in the global estimate in (3.38) results in them canceling out in the local estimate.

3.2.4   THE PERTURBED VECTOR FIELD AND FLOW

Consider the vector field

$$\dot{x} = f^{\text{pert}}(x), \qquad x \in \mathbb{R}^n, \tag{3.46}$$

and let $\mathcal{K} \supset M_2$ be a compact set, where $\mathcal{K}$ is the closure of an open set in $\mathbb{R}^n$ that contains M_2. Then we say that the vector field $f(x)$ is C^1 θ-close to $f^{\text{pert}}(x)$ if

$$\sup_{x \in \mathcal{K}} \| f(x) - f^{\text{pert}}(x) \| \leq \theta,$$

$$\sup_{x \in \mathcal{K}} \| Df(x) - Df^{\text{pert}}(x) \| \leq \theta. \tag{3.47}$$

We will refer to $f(x)$ as the *unperturbed vector field* and to $f^{\text{pert}}(x)$ as the *perturbed vector field* and we will be interested in comparing the dynamics of the time-T maps generated by the flows of the perturbed and unperturbed vector fields. The following lemma will provide adequate estimates.

Lemma 3.2.7 *Let $\phi_t(\cdot)$ and $\phi_t^{\text{pert}}(\cdot)$ denote the flows generated by the unperturbed and perturbed vector fields, respectively, and let $T > 1$ be fixed. Then for each $x \in \mathcal{K}$*

$$\| \phi_t(x) - \phi_t^{\text{pert}}(x) \| = \mathcal{O}\left(\theta T e^{LT}\right),$$

$$\| D\phi_t(x) - D\phi_t^{\text{pert}}(x) \| = \mathcal{O}\left(\theta T e^{2L_1 T}\right) \tag{3.48}$$

for all $|t| \leq T$, provided $\phi_t(x)$ and $\phi_t^{\text{pert}}(x)$ remain in the compact set $\mathcal{K}$ for $|t| \leq T$. The constant L is the Lipschitz constant of $f(x)$ and L_1 is the bound of $Df(x)$ on the compact set $\mathcal{K}$.

Proof: We write the perturbed and unperturbed vector fields in integral form and subtract them to obtain

$$\phi_t(x) - \phi_t^{\text{pert}}(x) = \phi_0(x) - \phi_0^{\text{pert}}(x) + \int_0^t [f(\phi_s(x)) - f^{\text{pert}}(\phi_s^{\text{pert}}(x))]\, ds. \tag{3.49}$$

For the purposes of our comparison we have $\phi_0(x) = \phi_0^{\text{pert}}(x) = x$. Hence,

taking norms of (3.49), using (3.47), and applying Gronwall's inequality gives the first result of the lemma. The second result for the linear flow is obtained in a similar way. $\qquad\square$

The following is a simple corollary of this lemma.

Corollary 3.2.8 *For $f(x)$ C^1 θ - close to $f^{\text{pert}}(x)$, and θ sufficiently small, we have*

$$\phi_T^{\text{pert}}\left(N_\epsilon'|_{\phi_{-T}(\overline{U}_i^4)}\right) \subset N_\epsilon'|_{U_i^5}.$$

Proof: This follows immediately from Lemmas 3.2.6 and 3.2.7.   □

As for the unperturbed time-T map, we can find local coordinate expressions for the perturbed time-T map as follows:

$$(x, y) \mapsto (f_{ij}(x, y), g_{ij}(x, y)),$$

where

$$
\begin{aligned}
f_{ij}(x, y) &= \sigma_j \circ \phi_T^{\text{pert}} \circ (\sigma_i \times \tau_i)^{-1}(x, y), & (3.50)\\
g_{ij}(x, y) &= \tau_j \circ \phi_T^{\text{pert}} \circ (\sigma_i \times \tau_i)^{-1}(x, y). & (3.51)
\end{aligned}
$$

As in the unperturbed case, these expressions are defined for

$$(x, y) \in (\sigma_i \times \tau_i) \, N'_\epsilon|_{\overline{U}_i^4 \cap \phi_{-T}(\overline{U}_j^4)}$$

for ϵ and θ sufficiently small (this follows from Corollary 3.2.8).

We collect some necessary estimates together in the following lemma.

Lemma 3.2.9 *Let $\eta > 0$ be given. Then for $f(x)$ C^1 θ - close to $f^{\text{pert}}(x)$ and θ and ϵ sufficiently small, we have*

$$\| (D_1 f_{ij}(x, y))^{-1} \|^k \| D_2 g_{ij}(x, y) \| < \tfrac{1}{2}, \quad 0 \le k \le r, \qquad (3.52)$$

$$\| g_{ij}(x, y) \| < \epsilon, \qquad (3.53)$$

$$\| D_1 g_{ij}(x, y) \| < \eta. \qquad (3.54)$$

Moreover, the norms of all first partial derivatives of f_{ij}^0, g_{ij}^0, f_{ij}, g_{ij}, as well as $(D_1 f_{ij})^{-1}$, are bounded on $(\sigma_i \times \tau_i) \, N'_\epsilon|_{\overline{U}_i^4 \cap \phi_{-T}(\overline{U}_j^4)}$, for all i, j by some constant, say Q.

Proof: Inequality (3.52) follows from (3.44) and Lemma 3.2.7, (3.53) follows from (3.45) and Lemma 3.2.7, and (3.54) follows from the fact that $g_{ij}^0(x, 0) = 0$. The latter part of the lemma follows immediately from the fact that we are dealing with continuous functions on compact sets.   □

It is important to emphasize that the closeness of $f(x)$ and $f^{\text{pert}}(x)$ is measured on a compact set that contains M. This is important since the phase space $\mathbb{R}^n$ is unbounded and, consequently, $f(x)$ and $f^{\text{pert}}(x)$ may become unbounded.

3.2.5 SMALLNESS AND CLOSENESS PARAMETERS

At this point we want to highlight the fact that the parameters ϵ, η and θ are controlling the issue of "smallness" and "closeness" in various arguments. The parameter ϵ controls the "thickness" of a neighborhood of M and, hence, the deviation of trajectories near M from trajectories of the dynamics linearized about M. The parameter θ controls the closeness of the unperturbed and perturbed dynamics. The parameter η can be made small, by taking ϵ and θ small enough, as a result of the overflowing invariance of M.

3.2.6 THE SPACE OF SECTIONS OF THE TRANSVERSAL BUNDLE

Let S denote the space of sections of $N'_\epsilon|_{\cup_{i=1}^s U_i^3}$. For any $u \in S$ we construct a coordinate representation, $u_i : \mathcal{D}^3 \longrightarrow \mathbb{R}^{n-k}$, defined by $u_i = \tau_i \circ u \circ \sigma_i^{-1}$. Thus, in coordinates, the section u is represented by the s maps u_i, $i = 1, \ldots, s$. The image in $\mathbb{R}^n$ of any $u \in S$ is a manifold having the same dimension as M. Near any coordinate chart domain, U_i^3, the image of u is represented as the graph of u_i. Inspired by this geometrical picture, we will henceforth refer to the image of a section u in $\mathbb{R}^n$ as $graph\ u$ or as a $graph$ $over\ M$. In Fig. 3.6 we give a geometrical illustration of these ideas.

We define

$$\mathrm{Lip}\, u \equiv \max_i \sup_{\substack{x,x' \in \mathcal{D}^3 \\ x \neq x'}} \frac{\| u_i(x) - u_i(x') \|}{\| x - x' \|}$$

if this exists, and denote the space of sections with Lipschitz constant δ by

$$S_\delta = \{u \in S : \mathrm{Lip}\, u \leq \delta\}.$$

The image in $\mathbb{R}^n$ of any $u \in S_\delta$ is a Lipschitz continuous manifold having the same dimension as M. Note that if $\delta = 0$, then in each local coordinate chart, u_i is an $n - k$-dimensional plane parallel to the x plane.

3.2.7 THE GRAPH TRANSFORM

We now define the $graph\ transform$. This will be a map of S_δ into S_δ that is constructed from the dynamics in such a way that a fixed point of this map is the overflowing invariant manifold for the perturbed vector field. To begin our construction we give a version of the implicit function theorem which will play a key role in showing that the graph transform is well defined.

Theorem 3.2.10 *Let $i(x) = x$ denote the inclusion map from $\mathcal{D}^4$ into $\mathbb{R}^{n-k}$. There is a neighborhood W of i in the space of Lipschitz maps from $\mathcal{D}^4$ into $\mathbb{R}^{n-k}$ equipped with the C^0 topology, such that for all $\chi \in W$:*

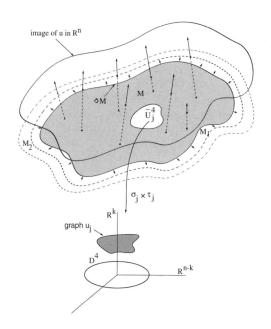

FIGURE 3.6. A Section of the Transversal Bundle.

1. χ is one-to-one.

2. $\overline{\mathcal{D}}^3 \subset \chi(\mathcal{D}^4) \subset \overline{\chi(\mathcal{D}^4)} \subset \mathcal{D}^5$.

Proof: See Dieudonné [1960], pp. 259-266.   □

We will use this result to prove the following proposition.

Proposition 3.2.11 *Let* $\omega : N'_\epsilon \longrightarrow M_1$ *denote the fiber projection map.*
We define the map $\chi(p) \equiv \omega \, \phi_T^{\mathrm{pert}} \, u \, \phi_{-T}(p)$. *Then there is a* θ *sufficiently*
small and $\delta > 0$, *such that for all* $f^{\mathrm{pert}}(x)$ C^1 θ-close to $f(x)$ and $u \in S_\delta$:

   1. $\chi(p)$ *is defined for all* $p \in \bigcup_{i=1}^s U_i^4$.

   2. $\overline{\bigcup_{i=1}^s U_i^3} \subset \chi\left(\bigcup_{i=1}^s U_i^4\right) \subset \overline{\chi\left(\bigcup_{i=1}^s U_i^4\right)} \subset \bigcup_{i=1}^s U_i^5$.

   3. *Each point in* $\bigcup_{i=1}^s U_i^3$ *is the* χ *image of only one point of* $\bigcup_{i=1}^s U_i^4$.

Proof: First note that for $\theta = 0$ and $\epsilon = 0$ the local representative of
$\chi(\cdot)$ becomes the inclusion map from $\mathcal{D}^4$ into $\mathbb{R}^{n-k}$. Thus, for θ and ϵ

sufficiently small the results of Theorem 3.2.10 will hold and can be applied in this context.

We begin by proving statement 1. Since, $T > 1$, for any $p \in \bigcup_{i=1}^{s} U_i^4 \subset M_1$, it follows that $\phi_{-T}(p) \in M \subset \bigcup_{i=1}^{s} U_i^3$. Therefore, the point $\phi_{-T}(p)$ lies in the domain of definition of the section u. Now $\| u(\phi_{-T}(p)) \| < \epsilon$; so by Corollary 3.2.8, $\| \phi_T^{\text{pert}} u(\phi_{-T}(p)) \| < \epsilon$; hence, $\omega \phi_T^{\text{pert}} u \phi_{-T}(p)$ is defined. Next we turn to statements 2 and 3.

For ϵ, θ, and δ sufficiently small, $\chi(\cdot)$ can be made as close to the identity map as we desire, and, thus, the local representative of χ can be made as close to the inclusion $i : \mathcal{D}^4 \to \mathbb{R}^{n-k}$ as we want. Hence, Theorem 3.2.10 applies, and statements 2 and 3 of this proposition are merely a restatement of results 2 and 1, respectively, of Theorem 3.2.10.   □

We now give a more geometrical interpretation of the map $\chi(\cdot)$ and the implications of statement 3 of Proposition 3.2.11. The main implication of this proposition is that if $u \in S_\delta$ (so that it is a section that is a graph over M), then the image of u under the time-T map of the perturbed flow is also a graph over M; in other words, the image of the section u cannot "fold over" and become "multivalued" (of course, we are restricting the domain of ϕ_T^{pert} (graph u) when viewed as a section, to M).

Let us make these ideas more precise. Suppose p and p' are two distinct points in M and let $u \in S_\delta$. Then $u\phi_{-T}(p)$ and $u\phi_{-T}(p')$ are two distinct points in graph u. Next we consider ϕ_T^{pert} (graph u) and, in particular, we consider the fate of the two points $u\phi_{-T}(p)$ and $u\phi_{-T}(p')$. Now if $\phi_T^{\text{pert}} u \phi_{-T}(p)$ and $\phi_T^{\text{pert}} u \phi_{-T}(p')$ are on the same fiber (by uniqueness of solutions they are distinct points), then $\omega \phi_T^{\text{pert}} u \phi_{-T}(p) \equiv \chi(p) = \omega \phi_T^{\text{pert}} u \phi_{-T}(p') \equiv \chi(p')$. Geometrically, this means that ϕ_T^{pert} (graph u) is not a section of $N_\epsilon'|_M$. Conversely, if $\chi(p) = \chi(p')$, then clearly ϕ_T^{pert} is not a graph when viewed as a section of $N_\epsilon'|_M$. Thus, Proposition 3.2.11 implies that for θ, ϵ, and δ sufficiently small, ϕ_T^{pert} (graph u) is a graph when viewed as a section of $N_\epsilon'|_M$. See Fig. 3.7 for an illustration of the geometry.

Thus, for ϵ, θ, and δ sufficiently small, $\phi_T^{\text{pert}}(\cdot)$ defines a map $G : S_\delta \longrightarrow S$ which we now describe in coordinates. Consider the map

$$(x, u_i(x)) \mapsto (f_{ij}(x, u_i(x)), g_{ij}(x, u_i(x))), \qquad (3.55)$$

which is defined for

$$(x, y) \in (\sigma_i \times \tau_i) \, N_\epsilon'|_{\overline{U_i^4} \cap \phi_{-T}(\overline{U_j^4})}.$$

In words, (3.55) describes the image of a point on graph u_i under the time-T map of the perturbed flow, and the domain of definition of this map ensures that the image of this point under $\phi_T^{\text{pert}} \circ (\sigma_i \times \tau_i)^{-1}$ lies on a fiber over a subset of $\overline{U_j^5}$. By Proposition 3.2.11, the point $(f_{ij}(x, u_i(x)), g_{ij}(x, u_i(x)))$

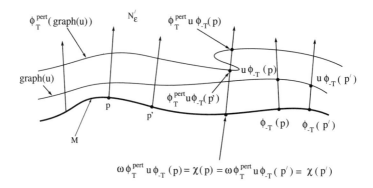

FIGURE 3.7. A Geometrical Illustration of Proposition 3.2.5.

lies on graph $\phi_T^{\mathrm{pert}}(u)$. In the coordinate chart domain U_i^3 we represent graph $\phi_T^{\mathrm{pert}}(u)$ as

$$(Gu)_j \left(f_{ij}(x, u_i(x)) \right) = g_{ij}(x, u_i(x)). \tag{3.56}$$

The condition for graph u to be invariant under the time-T map of the perturbed flow, expressed in coordinates, is

$$u_j \left(f_{ij}(x, u_i(x)) \right) = g_{ij}(x, u_i(x)), \tag{3.57}$$

which is also the condition for u to be a fixed point of G. We will show that $G : S_\delta \longrightarrow S$ has a unique fixed point, say u. After this is established, we will show that graph u is overflowing invariant under $\phi_t^{\mathrm{pert}}(\cdot)$ for $t > 0$. The map G is referred to as the *graph transform*. We illustrate the geometry of the graph transform in Fig. 3.8.

> **Technical Detail:** We have argued that the image of a section of $N_\epsilon'|_M$ under the graph transform is still a graph, i.e., a section. However, we need to be aware of the domain of this section. Under iteration by G, a section may *overflow* the boundary of $\bar{M}$. In this case we "cutoff" this part of the section by restricting the section to the fibers defined over M, i.e., we will always consider $Gu|_{N_\epsilon'|_M}$.

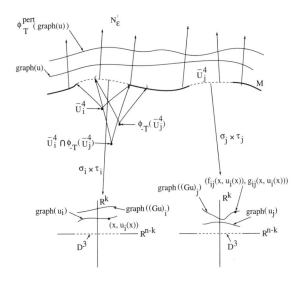

FIGURE 3.8. Geometry of the Graph Transform.

3.3   Statement and Proof of the Main Theorem

We can now state and prove Fenichel's theorem on the persistence of over-flowing invariant manifolds.

Theorem 3.3.1 (Fenichel, 1971) *Suppose $\dot{x} = f(x)$ is a C^r vector field on $\mathbb{R}^n$, $r \geq 1$. Let $\bar{M} \equiv M \cup \partial M$ be a C^r compact connected manifold with boundary overflowing invariant under the vector field $f(x)$. Suppose $\nu(p) < 1$ and $\sigma(p) < \frac{1}{r}$ for all $p \in M$. Then for any C^r vector field $f^{\mathrm{pert}}(x)$ C^1 θ-close to $f(x)$, with θ sufficiently small, there is a manifold $\bar{M}^{\mathrm{pert}}$ overflowing invariant under $f^{\mathrm{pert}}(x)$ and C^r diffeomorphic to $\bar{M}$.*

The proof is divided into two distinct pieces. First we prove the existence of a Lipschitz manifold that is overflowing invariant under $f^{\mathrm{pert}}(x)$. After this we will show that under the hypotheses of the theorem this Lipschitz manifold is actually C^r.

We begin with a proposition.

Proposition 3.3.2 *For ϵ, δ, and η sufficiently small, $G : S_\delta \longrightarrow S_\delta$.*

Proof: We will show that for $u \in S_\delta$

$$\| (Gu)_j \, (\xi) - (Gu)_j \, (\xi') \| \leq \delta \, \| \xi - \xi' \| \tag{3.58}$$

for all ξ, $\xi' \in \mathcal{D}^3$ and for each j. (Recall that the Lipschitz constant was was defined in the local coordinate charts.) In fact, it is sufficient to show that this inequality holds for all ξ' in a neighborhood of a given ξ. The reason that this is sufficient is that for *any* ξ' and ξ in $\mathcal{D}^3$ the line segment joining ξ' and ξ also lies in $\mathcal{D}^3$ (because $\mathcal{D}^3$ is convex) and this line can be covered by a finite number of such neighborhoods. Hence, the Lipschitz inequality will hold on each such neighborhood and, therefore, throughout $\mathcal{D}^3$.

Now choose any $p \in U_j^3$. Then by part 2 of Proposition 3.2.11 there is point p_- in some U_i^3 such that $\omega\phi_T^{\mathrm{pert}}u(p_-) = p$. Let $x = \sigma_i(p_-)$ and choose $x' \in \mathcal{D}^3$ near x. Then the graph transform in local coordinates is defined and is expressed as

$$(Gu)_j(\xi) = g_{ij}(x, u_i(x))$$

and

$$(Gu)_j(\xi') = g_{ij}(x', u_i(x')),$$

where

$$\xi = f_{ij}(x, u_i(x))$$

and

$$\xi' = f_{ij}(x', u_i(x')).$$

Technical Note: Several times throughout this proof the phrase *for x sufficiently close to x'* is used. The justification for this is related to the remark at the very beginning of the proof of the proposition that it is sufficient to show (3.58) for all ξ sufficiently close to a given ξ'. Taking x close to x' forces ξ to be close to ξ' since $\xi = f_{ij}(x, u_i(x))$ and $\xi' = f_{ij}(x', u_i(x'))$. Thus, we are free to apply arguments that require x to be close to x'.

We introduce the following shorthand notation:

$$
\begin{align}
A &\equiv D_1 g_{ij}(x, u_i(x)), \tag{3.59}\\
B &\equiv D_2 g_{ij}(x, u_i(x)), \tag{3.60}\\
C &\equiv D_1 f_{ij}(x, u_i(x)), \tag{3.61}\\
E &\equiv D_2 f_{ij}(x, u_i(x)), \tag{3.62}
\end{align}
$$

where A, B, C, and E depend on x, i, j, u, and $f^{\mathrm{pert}}(x)$. Translating the results of Lemma 3.2.9 into this shorthand notation, we have the following bounds:

$$
\begin{align}
&\| A \|, \| B \|, \| C \|, \| C^{-1} \|, \| E \| \le Q, \tag{3.63}\\
&\| A \| < \eta, \tag{3.64}\\
&\| B \| < \tfrac{1}{2}, \tag{3.65}\\
&\| B \| \| C^{-1} \| < \tfrac{1}{2}. \tag{3.66}
\end{align}
$$

As a preliminary estimate we need to obtain a lower bound on $\| \xi - \xi' \|$ in terms of $\| x - x' \|$. Toward this end, we have

$$\| \xi - \xi' \| = \| f_{ij}(x, u_i(x)) - f_{ij}(x', u_i(x')) \|$$

$$\geq \underbrace{\| f_{ij}(x, u_i(x)) - f_{ij}(x', u_i(x)) \|}_{\text{obtain lower bound}} \qquad (3.67)$$

$$- \underbrace{\| f_{ij}(x', u_i(x)) - f_{ij}(x', u_i(x')) \|}_{\text{obtain upper bound}}.$$

We next obtain a lower bound for the first term and an upper bound for the second term of (3.67) as indicated. Obtaining the lower bound for the first term requires certain inversions that must be justified via the implicit function theorem.

Technical Detail: The Lower Bound for $\| f_{ij}(x, u_i(x)) - f_{ij} (x', u_i(x)) \|$: We use a general notation for these details for the sake of clarity. Consider the two equations

$$z_1 = F(x_1, y),$$
$$z_2 = F(x_2, y),$$

where $x_i, z_i \in \mathbb{R}^{n-k}$, $i = 1, 2$, and $y \in \mathbb{R}^k$ is regarded as a parameter vector. The function F is sufficiently differentiable (say C^r, $r \geq 2$) on an open set where the implicit function theorem can be applied.

Using the implicit function theorem, we invert these two equations to obtain

$$x_1 = H(z_1, y),$$
$$x_2 = H(z_2, y).$$

Taylor expanding and subtracting these equations gives

$$x_1 - x_2 = DH (z_1 - z_2) + \mathcal{O} \left(\| z_1 - z_2 \|^2 \right),$$

or

$$x_1 - x_2 = DF^{-1} (F(x_1, y) - F(x_2, y)) + \mathcal{O} \left(\| x_1 - x_2 \|^2 \right).$$

Translating these estimates into our notation by letting $F = f_{ij}$, $x_1 = x$, $x_2 = x'$, $y = u_i(x)$, $C^{-1} = DF^{-1}$, we obtain

$$\| f_{ij}(x, u_i(x)) - f_{ij}(x', u_i(x)) \| \geq \| C^{-1} \|^{-1} \| x - x' \| + \mathcal{O}\left(\| x - x' \|^2 \right).$$
(3.68)

Obtaining an upper bound for the second term of (3.67) is trivial. Taylor expanding this term gives

$$\| f_{ij}(x', u_i(x)) - f_{ij}(x', u_i(x')) \|$$

$$\leq \| E \| \| u_i(x) - u_i(x') \| + \mathcal{O}\left(\| u_i(x) - u_i(x') \|^2 \right)$$

$$\leq \delta \| E \| \| x - x' \| + \mathcal{O}\left(\| x - x' \|^2 \right).$$
(3.69)

Using (3.67), (3.68), and (3.69), we obtain

$$\| \xi - \xi' \| \geq \| C^{-1} \|^{-1} \| x - x' \| - \delta \| E \| \| x - x' \| + \mathcal{O}\left(\| x - x' \|^2 \right)$$

$$\geq \| C^{-1} \|^{-1} (1 - \delta \| C \| \| E \|) \| x - x' \| + \mathcal{O}\left(\| x - x' \|^2 \right).$$
(3.70)

Now

$$\| C \| \| E \| < Q^2, \qquad \| C^{-1} \| < Q,$$

and we choose x, x' suficiently close so that

$$\mathcal{O}\left(\| x - x' \|^2 \right) < \delta Q \| x - x' \|.$$

Combining these two estimates with (3.70) gives

$$\| \xi - \xi' \| \geq \| C^{-1} \|^{-1} \left(1 - \delta Q^2\right) \| x - x' \| - \delta Q \| C^{-1} \| \| C^{-1} \|^{-1} \| x - x' \|$$

or

$$\| \xi - \xi' \| \geq \| C^{-1} \|^{-1} \left(1 - 2\delta Q^2\right) \| x - x' \|.$$
(3.71)

Now we return to the main estimate

$$\| (Gu)_j(\xi) - (Gu)_j(\xi') \| = \| g_{ij}(x, u_i(x)) - g_{ij}(x', u_i(x')) \|$$

$$\leq \| g_{ij}(x, u_i(x)) - g_{ij}(x', u_i(x)) \|$$

$$+ \| g_{ij}(x', u_i(x)) - g_{ij}(x', u_i(x')) \|. \quad (3.72)$$

We Taylor expand each of the two terms in the second line of (3.72) to obtain

$$\| g_{ij}(x, u_i(x)) - g_{ij}(x', u_i(x)) \| \leq \| A \| \| x - x' \| + \mathcal{O}\left(\| x - x' \|^2\right),$$

$$\| g_{ij}(x', u_i(x)) - g_{ij}(x', u_i(x')) \| \leq \| B \| \delta \| x - x' \| + \mathcal{O}\left(\| x - x' \|^2\right). \tag{3.73}$$

Substituting (3.73) into (3.72) gives

$$\| (Gu)_j(\xi) - (Gu)_j(\xi') \| \leq (\| A \| + \| B \| \delta) \| x - x' \| + \mathcal{O}\left(\| x - x' \|^2\right). \tag{3.74}$$

Now we use the bound $\| A \| < \eta$ and we choose x and x' sufficiently close so that

$$\mathcal{O}\left(\| x - x' \|^2\right) < \eta \| x - x' \|.$$

Using these estimates, (3.74) becomes

$$\begin{aligned} \| (Gu)_j(\xi) - (Gu)_j(\xi') \| &\leq (2\eta + \| B \| \delta) \| x - x' \| \\ &\leq \frac{(2\eta + \| B \| \delta)}{1 - 2\delta Q^2} \| C^{-1} \| \| \xi - \xi' \|, \end{aligned} \tag{3.75}$$

where we have used (3.71) in passing from the first to the second line of (3.75). We choose δ sufficiently small such that

$$1 - 2\delta Q^2 > \frac{3}{4},$$

and η sufficiently small such that

$$2\eta Q < \frac{2\delta}{8}.$$

Then since $\| B \| \| C^{-1} \| < \frac{1}{2}$ we have

$$\frac{(2\eta + \| B \| \delta)}{1 - 2\delta Q^2} \| C^{-1} \| < \frac{\frac{2\delta}{8} + \frac{\delta}{2}}{\frac{3}{4}} = \delta. \tag{3.76}$$

Combining (3.76) and (3.75) gives the result.   □

Proposition 3.3.3 *For ϵ, δ, and η sufficiently small G is a contraction on S_δ in the C^0 norm.*

Proof: Let u, $u' \in S_\delta$ and let $\xi \in \sigma_j(U_j^3)$ be given. Choose i, x, x' such that

$$\xi = f_{ij}(x, u_i(x)) = f_{ij}(x', u_i'(x')). \tag{3.77}$$

We have the following basic estimate which we will subsequently refine:

$$\| (Gu)_j(\xi) - (Gu')_j(\xi) \| = \| g_{ij}(x, u_i(x)) - g_{ij}(x', u_i'(x')) \|$$

$$\leq \quad \| g_{ij}(x, u_i(x)) - g_{ij}(x', u_i(x)) \| + \| g_{ij}(x', u_i(x)) - g_{ij}(x', u_i'(x)) \|$$

$$+ \| g_{ij}(x', u_i'(x)) - g_{ij}(x', u_i'(x')) \|$$

$$\leq \quad \| A \| \| x - x' \| + \| B \| \| u_i(x) - u_i'(x) \|$$

$$+ \| B \| \| u_i'(x) - u_i'(x') \| + \mathcal{O}\left(\| x - x' \|^2 \right)$$

$$+ \mathcal{O}\left(\| u_i(x) - u_i'(x) \|^2 \right) + \mathcal{O}\left(\| u_i'(x) - u_i'(x') \|^2 \right)$$

$$\leq \quad (2\eta + \tfrac{1}{2}\delta) \| x - x' \| + \frac{1}{2} \| u_i - u_i' \|_0$$

$$+ \mathcal{O}\left(\| u_i - u_i' \|_0^2 \right). \tag{3.78}$$

First, recall that $\| u \| < \epsilon$; hence, for ϵ sufficiently small we have

$$\mathcal{O}\left(\| u_i - u_i' \|_0^2 \right) < \frac{1}{16} \| u_i - u_i' \|_0. \tag{3.79}$$

Next, we need to estimate $\| x - x' \|$ in terms of $\| u_i - u_i' \|_0$. This will involve two steps. From (3.68) we have

$$\| f_{ij}(x, u_i(x)) - f_{ij}(x', u_i(x)) \| \geq \| C^{-1} \|^{-1} \| x - x' \| + \mathcal{O}\left(\| x - x' \|^2 \right)$$

$$\geq \tfrac{1}{2} \| C^{-1} \|^{-1} \| x - x' \| \tag{3.80}$$

for x and x' such that

$$|\mathcal{O}\left(\| x - x' \|^2 \right)| < \tfrac{1}{2} \| C^{-1} \|^{-1} \| x - x' \|.$$

Next we estimate $\| f_{ij}(x', u_i'(x')) - f_{ij}(x', u_i(x)) \|$ in terms of $\| x - x' \|$ and $\| u_i - u_i' \|_0$. To this end we have

$$\| f_{ij}(x', u_i'(x')) - f_{ij}(x', u_i(x)) \|$$

$$\leq \ \parallel f_{ij}(x', u_i'(x')) - f_{ij}(x', u_i'(x)) \parallel + \parallel f_{ij}(x', u_i'(x)) - f_{ij}(x', u_i(x)) \parallel$$

$$\leq \ \parallel E \parallel \parallel u_i'(x') - u_i'(x)) \parallel + \mathcal{O}\left(\parallel u_i'(x') - u_i'(x) \parallel^2 \right)$$

$$+ \parallel E \parallel \parallel u_i'(x) - u_i(x)) \parallel + \mathcal{O}\left(\parallel u_i'(x) - u_i(x) \parallel^2 \right)$$

$$\leq \ Q\left(\delta \parallel x - x' \parallel + \parallel u_i' - u_i \parallel_0 \right) + \mathcal{O}\left(\parallel x - x' \parallel^2 \right) + \mathcal{O}\left(\parallel u_i' - u_i \parallel_0^2 \right).$$

$$(3.81)$$

Combining (3.80) and (3.81), and using (3.77), gives

$$\begin{aligned}
\tfrac{1}{2} \parallel C^{-1} \parallel^{-1} \parallel x - x' \parallel \ \leq \ & Q\delta \parallel x - x' \parallel + Q \parallel u_i' - u_i \parallel \\
& + \mathcal{O}\left(\parallel x - x' \parallel^2 \right) + \mathcal{O}\left(\parallel u_i'(x) - u_i(x) \parallel^2 \right),
\end{aligned}$$

from which, for δ sufficiently small, it follows that

$$\parallel x - x' \parallel < \text{constant} \parallel u_i - u_i' \parallel_0 .$$

Combining this estimate with (3.78) and (3.79) gives

$$\parallel (Gu)_j (\xi) - (Gu')_j (\xi') \parallel \leq \left[\left(2\eta + \frac{1}{2}\delta \right) \text{constant} + \frac{9}{16} \right] \parallel u_i - u_i' \parallel_0 .$$

$$(3.82)$$

For δ and η sufficiently small, it follows that

$$\left(2\eta + \frac{1}{2}\delta \right) \text{constant} + \frac{9}{16} < 1.$$

Hence, the proposition is proved. □

Corollary 3.3.4 *There is a unique $u \in S_\delta$ such that $\phi_t^{\mathrm{pert}} (\mathrm{graph}\, u) \supset$
graph u for all $t > 0$.*

Proof: S_δ is a metric space closed under C^0 convergence; hence, by the
contraction mapping principle, G has a unique fixed point, which we call
u. From an argument similar to Proposition 3.2.11 it follows that for small
$t > 0$ $\phi_t^{\mathrm{pert}} (\mathrm{graph}\, u) \cap N_\epsilon'|_{\cup_{i=1}^s U_i^3}$ is the graph of an element $u_t \in S_\delta$. Since
u is a fixed point of G, we have

$$\mathrm{graph}\, u \subset \phi_T^{\mathrm{pert}} (\mathrm{graph}\, u)$$

and, therefore,

$$\phi_t^{\text{pert}}\left(\text{graph}\,u\right) \subset \phi_t^{\text{pert}}\left(\phi_T^{\text{pert}}\left(\text{graph}\,u\right)\right) = \phi_T^{\text{pert}}\left(\phi_t^{\text{pert}}\left(\text{graph}\,u\right)\right).$$

Combining this expression with the fact that the fixed point of G is unique allows us to conclude that

$$u_t = u.$$

This argument can be repeated indefinitely to give the result.   □

3.3.1   DIFFERENTIABILITY OF THE PERSISTING OVERFLOWING INVARIANT MANIFOLD

The contraction mapping argument only gives the existence of an overflowing invariant *Lipschitz* manifold for the perturbed vector field. We must now take this manifold and work harder in order to show that it is actually C^r under the hypotheses of Theorem 3.3.1.

Let u denote the section of the transversal bundle whose image in $\mathbb{R}^n$ is the persisting overflowing invariant manifold. Then, in the local coordinate charts, u is represented by s maps as

$$\tau_i u \sigma_i^{-1} \equiv u_i\,:\,\mathcal{D}^3 \longrightarrow \mathbb{R}^k. \tag{3.83}$$

If u_i is C^1, then the derivative of u_i, denoted Du_i, assigns to each point in $\mathcal{D}^3$ a linear map from $\mathbb{R}^{n-k}$ to $\mathbb{R}^k$. Thus, the derivative of the section u, Du, is represented in the local cordinate charts by s maps

$$Du_i \equiv v_i \in C^0\left(\mathcal{D}^3, \mathrm{L}\left(\mathbb{R}^{n-k}, \mathbb{R}^k\right)\right). \tag{3.84}$$

Accordingly, we denote Du by the following s-vector:

$$v = (v_1, \ldots, v_s) \in \left[C^0\left(\mathcal{D}^3, \mathrm{L}\left(\mathbb{R}^{n-k}, \mathbb{R}^k\right)\right)\right]^s \tag{3.85}$$

and define the norm of v as

$$\| \, v \, \| \equiv \max_i \sup_{x \in \mathcal{D}^3} \| \, v_i(x) \, \| . \tag{3.86}$$

With this norm, $[C^0(\mathcal{D}^3, \mathrm{L}(\mathbb{R}^{n-k}, \mathbb{R}^k))]^s$ is a complete metric space having the topology of uniform convergence.

With these preliminaries out of the way, our strategy for proving that u is C^r is as follows:

1. First prove that $u \in C^1$. This will be accomplished in the following steps:

(a) Derive a formal equation that Du must satisfy if it were differentiable.

(b) Use iteration and a contraction mapping argument to show that this equation has a solution.

(c) Show that the solution thus obtained is, indeed, the derivative of u.

2. Use the same type of arguments inductively to show that u is C^r.

We now obtain a *formal* equation that the derivative must satisfy if it exists. Recall from (3.57) that in the local coordinate charts, u must satisfy

$$u_j(\xi) = g_{ij}(x, u_i(x)),\tag{3.87}$$

where

$$\xi = f_{ij}(x, u_i(x)).\tag{3.88}$$

Differentiating these expressions formally gives

$$(Du_j)\, D\xi = D_1 g_{ij}(x, u_i(x)) + D_2 g_{ij}(x, u_i(x)) Du_i(x)\tag{3.89}$$

and

$$D\xi = D_1 f_{ij}(x, u_i(x)) + D_2 f_{ij}(x, u_i(x)) Du_i(x).\tag{3.90}$$

Combining these two expressions gives

$$
\begin{aligned}
Du_j(\xi) = {} & [D_1 g_{ij}(x, u_i(x)) + D_2 g_{ij}(x, u_i(x)) Du_i(x)] \\
& \times [D_1 f_{ij}(x, u_i(x)) + D_2 f_{ij}(x, u_i(x)) Du_i(x)]^{-1}
\end{aligned}\tag{3.91}
$$

or, recalling the shorthand notation given in the proof of Proposition 3.3.2,

$$
\begin{aligned}
A &\equiv D_1 g_{ij}(x, u_i(x)), \\
B &\equiv D_2 g_{ij}(x, u_i(x)), \\
C &\equiv D_1 f_{ij}(x, u_i(x)), \\
E &\equiv D_2 f_{ij}(x, u_i(x)),
\end{aligned}
$$

where A, B, C, and E depend on x, i, j, u, and $f^{\mathrm{pert}}(x)$. Equation (3.91) can be written succinctly as

$$v_j(\xi) = [A + Bv_i(x)]\,[C + Ev_i(x)]^{-1}.\tag{3.92}$$

The components of the derivative of u must satisfy this functional equation. Introducing some additional shorthand notation,

$$H_{ij} v_i = [A + Bv_i(x)]\,[C + Ev_i(x)]^{-1}.\tag{3.93}$$

Hence, the functional equation which the derivative of u must satisfy has the form

$$v_j(\xi) = H_{ij}v_i(\xi). \tag{3.94}$$

At this point we are faced with an important issue; namely, the graph

transform could be formulated or discussed in a coordinate-free fashion. However, the H_{ij} do not have a coordinate-free meaning. Equation (3.94) is a relationship between quantities that are defined in two different coordinate charts. It also is dependent upon the global dynamics in the overflowing invariant section u.

We deal with this issue by combining the s functional equations into a single functional equation using a partition of unity; i.e., we choose a set of C^r functions

$$\phi_i : \bigcup_{i=1}^{s} U_i^3 \longrightarrow [0, 1] \tag{3.95}$$

with the support of ϕ_i contained in U_i^2 and $\sum_{i=1}^{s} \phi_i = 1$ on $\bigcup_{i=1}^{s} U_i^1$. (Note: Virtually any differential geometry textbook will prove that partitions of unity of this type exist on a manifold; see, e.g., Spivak [1979].) The derivative then satisfies the equation

$$v_j(\xi) = \sum_{i=1}^{s} \phi_i(p_-)H_{ij}v_i(\xi), \quad j = 1, \ldots, s, \tag{3.96}$$

which we solve via the following iteration scheme:

$$v_j^0(\xi) = 0,$$

$$v_j^{n+1}(\xi) = \sum_{i=1}^{s} \phi_i(p_-)H_{ij}v_i^n(\xi), \quad j = 1, \ldots, s, \tag{3.97}$$

where

$$\phi_T^{\text{pert}} u(p_-) = u\left(\sigma_j^{-1}(\xi)\right). \tag{3.98}$$

Also, note that $p_- \in M$ so that $\sum_{i=1}^{s} \phi_i(p_-) = 1$. It should be clear that the significance of the point p_- is that it is the point in U_i^2 of the unperturbed overflowing invariant manifold for which $\sigma_i(p_-) = x$.

We will show that (3.97) has a fixed point and that the fixed point is the derivative of u. This will be accomplished in two parts. First, we show the existence of a fixed point via an iteration method that is developed through a series of propositions. Second, we show that this fixed point is, indeed, the derivative of u by dealing directly with the definition of the derivative.

Proposition 3.3.5 *For ϵ, δ, and η sufficiently small, $\| v^n \| < \delta$ for each n.*

Proof: Upon examining the form of (3.97) and the definition of the norm of v given in (3.86) we see that it will be sufficient to prove that

$$\| H_{ij} v_i^{n-1}(\xi) \| < \delta \tag{3.99}$$

for all i, j, n and ξ such that

$$\omega \phi_{-T}^{\text{pert}} u \left(\sigma_j^{-1}(\xi) \right) \in U_i^3. \tag{3.100}$$

The proof is by induction. The proposition is obviously true for $n = 1$. Suppose it is true for $n = k$, i.e.,

$$\| v_j^k \| = \| H_{ij} v_i^{k-1} \| < \delta, \qquad \forall\, i,\, j. \tag{3.101}$$

We show that this implies that the proposition is true for $n = k + 1$.

From (3.93) and (3.97) we have

$$
\begin{aligned}
v_j^{k+1} &= H_{ij} v_i^k \\
&= \left[A + B v_i^k \right] \left[C + E v_i^k \right]^{-1} \\
&= \left[A + B v_i^k \right] \left[\mathbb{1} + E v_i^k C^{-1} \right]^{-1} C^{-1},
\end{aligned} \tag{3.102}
$$

where $\mathbb{1}$ is our general symbol for the identity operator. From this we obtain the estimate

$$\| v_j^{k+1} \| \leq \| A + B v_i^k \| \, \| C^{-1} \| \, \| \left[\mathbb{1} + E v_i^k C^{-1} \right]^{-1} \|. \tag{3.103}$$

By the induction hypothesis we have $\| v_i^k \| < \delta$ and we have the previous estimates (see the proof of Proposition 3.3.2)

$$
\begin{aligned}
&\| A \|,\, \| B \|,\, \| C \|,\, \| C^{-1} \|,\, \| E \| \leq Q, \\
&\| A \| < \eta, \\
&\| B \| < \tfrac{1}{2}, \\
&\| B \| \, \| C^{-1} \| < \tfrac{1}{2}.
\end{aligned}
$$

Using these estimates, it follows that $\| E v_i^k C^{-1} \| < Q^2 \delta$; hence,

$$\| \left[\mathbb{1} + E v_i^k C^{-1} \right]^{-1} \| < 1 + Q^2 \delta + \mathcal{O}(\delta^2).$$

Combining these results with (3.103) gives

$$
\begin{aligned}
\| v_j^{k+1} \| &\leq \left(\| A \| + \| B \| \delta \right) \| C^{-1} \| \left(1 + Q^2 \delta + \mathcal{O}(\delta^2) \right) \\
&\leq \left(Q\eta + \frac{\delta}{2} \right) \left(1 + Q^2 \delta + \mathcal{O}(\delta^2) \right) \\
&= \eta \left(Q + Q^3 \delta \right) + \frac{\delta}{2} + \mathcal{O}(\delta^2) + \mathcal{O}(\eta\delta) \\
&\leq \delta \qquad\qquad\qquad\qquad\qquad\qquad\qquad\qquad (3.104)
\end{aligned}
$$

for η sufficiently small. This completes the proof. □

We will use this result in the following proposition which is the key step in our iteration argument.

Proposition 3.3.6 *For ϵ, δ, and η sufficiently small, $\| v^{n+1} - v^n \| \leq \frac{3}{4} \| v^n - v^{n-1} \|$.*

Proof: The result follows from the following straightforward estimates.

$$
\| v_j^{n+1} - v_j^n \| = \| H_{ij} v_i^n - H_{ij} v_i^{n-1} \|
$$

$$
= \| [A + Bv_i^n] [C + Ev_i^n]^{-1} - [A + Bv_i^{n-1}] [C + Ev_i^{n-1}]^{-1} \|
$$

$$
= \| [A + Bv_i^n] [C + Ev_i^n]^{-1} \left([C + Ev_i^{n-1}] - [C + Ev_i^n] \right) [C + Ev_i^{n-1}]^{-1}
$$

$$
+ \left([A + Bv_i^n] - [A + Bv_i^{n-1}] \right) [C + Ev_i^{n-1}]^{-1} \|
$$

$$
\leq \delta \| E \| \| C^{-1} \| (1 + \mathcal{O}(\delta)) \| v_i^n - v_i^{n-1} \|
$$

$$
+ \| B \| \| C^{-1} \| (1 + \mathcal{O}(\delta)) \| v_i^n - v_i^{n-1} \|
$$

$$
\leq \left(\| B \| \| C^{-1} \| + \mathcal{O}(\delta) \right) \| v_i^n - v_i^{n-1} \|
$$

$$
\leq \frac{3}{4} \| v_i^n - v_i^{n-1} \|, \qquad\qquad\qquad\qquad\qquad (3.105)
$$

with the last inequality holding for η and δ sufficiently small. □

> **Technical Point:** In the estimates used in the proof of this corollary the small parameter η appears to be conspicuously absent. However, η does actually come into play as Proposition 3.3.5 is used to estimate the term $\| [A + Bv_i^n] [C + Ev_i^n]^{-1} \|$, and this proposition requires η to be taken sufficiently small. This will also be used in later estimates in this section.

Corollary 3.3.7 *The sequence v^n converges to a solution v, and the components of v satisfy the equation*

$$v_j = \sum_{i=1}^{s} \phi_i H_{ij} v_i.$$

Proof: From Proposition 3.3.6, $\{v^n\}$ is a Cauchy sequence. Since $[C^0(\mathcal{D}^3, \mathrm{L}(\mathbb{R}^{n-k}, \mathbb{R}^k))]^s$ is a complete metric space (with the metric obtained form the norm), then the Cauchy sequence converges to an element, say v, in this space. By construction, the components of v satisfy the functional equation.   □

The following lemma will play a key role in our establishing that v is, indeed, the derivative of u.

Lemma 3.3.8 *Suppose $\gamma(a)$ is a nondecreasing, non-negative function and satisfies the inequality*

$$\gamma(a) \le \alpha\gamma(\beta a) + r(a),$$

where a is small, $0 \le \alpha < 1$, and $\lim_{a \to 0} r(a) = 0$. Then

$$\gamma(a) \to 0 \qquad as \quad a \to 0.$$

Proof:
    There are two cases to be considered, $\beta \le 1$ and $\beta > 1$.

$\boxed{\beta \le 1}$

In this case we have

$$\gamma(a) \le \alpha\gamma(\beta a) + r(a) \le \alpha\gamma(a) + r(a)$$

or

$$\gamma(a) \le \frac{r(a)}{1-\alpha},$$

from which it immediately follows that $\gamma(a) \to 0$ as a $\to 0$. The second case is a bit more difficult.

$\boxed{\beta > 1}$

We begin by taking the inequality $\gamma(a) \le \alpha\gamma(\beta a) + r(a)$ and creating n new inequalities from it by replacing a successively by $a\beta^{-1}, a\beta^{-2}, \ldots, a\beta^n$

and weighting the terms with α^{n-1}, α^{n-2}, ..., 1. The inequalities obtained in this way are as follows:

$$
\begin{aligned}
\alpha^{n-1}\gamma(\alpha\beta^{-1}) &\leq \alpha^n\gamma(a) + \alpha^{n-1}r(\alpha\beta^{-1}), \\
\alpha^{n-2}\gamma(\alpha\beta^{-2}) &\leq \alpha^{n-1}\gamma(\alpha\beta^{-1}) + \alpha^{n-2}r(\alpha\beta^{-2}), \\
&\;\;\vdots \\
\alpha\gamma(\alpha\beta^{-(n-1)}) &\leq \alpha^2\gamma(\alpha\beta^{-(n-2)}) + \alpha r(\alpha\beta^{-(n-1)}), \\
\gamma(\alpha\beta^{-n}) &\leq \alpha\gamma(\alpha\beta^{-(n-1)}) + r(\alpha\beta^{-n}).
\end{aligned}
$$

Looking up from the bottom of these rows of inequalities, we see that the first term to the *right* of the $\leq$ sign is the term on the *left* of the $\leq$ sign on the row immediately above. This allows us to construct the following series of bounds on $\gamma(\alpha\beta^{-n})$:

$$
\begin{aligned}
\gamma(\alpha\beta^{-n}) &\leq \alpha^2\gamma(\alpha\beta^{-(n-2)}) + \alpha r(\alpha\beta^{-(n-1)}) + r(\alpha\beta^{-n}) \\
&\;\;\vdots \\
&\leq \alpha^n\gamma(a) + r(\alpha\beta^{-n}) + \alpha r(\alpha\beta^{-n+1}) + \cdots + \alpha^{n-1}r(\alpha\beta^{-1}) \\
&\leq \alpha^n\gamma(a) + \left(1 + \alpha + \cdots + \alpha^{n-1}\right)r(\alpha\beta^{-1}) \\
&\leq \alpha^n\gamma(a) + \frac{1}{1-\alpha}|r(\alpha\beta^{-1})|. && (3.106)
\end{aligned}
$$

The result follows from this last inequality. To see this, pass to the limit of both sides of (3.106) as $a \to 0$. Now $0 \leq \alpha < 1$ and n is arbitrary, from which it follows that we must have $\gamma(a) \to 0$ as $a \to 0$. $\qquad\square$

Proposition 3.3.9 *For all $p \in U_i^3$, $Du_i(\sigma_i(p))$ exists and is equal to $v_i(\sigma_i(p))$. Hence, u is C^1 and $v_j = H_{ij}v_i$.*

Proof: Motivated by the definition of the derivative, we define the increasing function

$$
\gamma : (0, 1) \longrightarrow \mathbb{R}, \tag{3.107}
$$

where

$$
\gamma(a) \equiv \max_i \; \sup_{\substack{\xi,\xi' \in \mathcal{D}^3 \\ 0 < \|\xi-\xi'\| < a}} \frac{\| u_i(\xi') - u_i(\xi) - v_i(\xi)(\xi - \xi') \|}{\| \xi' - \xi \|}. \tag{3.108}
$$

This function is nondecreasing because it is the supremum over a set that increases as a increases.

Using the fact that $u_i \in S_\delta$ and Proposition 3.3.5, we have immediately that $\gamma(a) < 2\delta$. To prove Proposition 3.3.9, it suffices to show that $\gamma(a) \to 0$ as $a \to 0$. To do this we will use Lemma 3.3.8. From this, it follows that

$$\lim_{\|\xi'-\xi\|\to 0} \frac{\| u_i(\xi') - u_i(\xi) - v_i(\xi)\,(\xi - \xi') \|}{\| \xi' - \xi \|} = 0$$

for every i, which implies that v_i s the derivative of u_i.

First, we address a technical issue. Let $\xi \in \mathcal{D}^3$ be given and suppose that for some i, j that $\xi = f_{ij}(x, u_i(x))$ with $x \in \mathcal{D}^3$. If d is chosen small enough and $\xi' \in \mathcal{D}^3$ with $\| \xi - \xi' \| < d$, there is an $x' \in \mathcal{D}^3$ such that $\xi' = f_{ij}(x', u_i(x'))$. Moreover, by compactness, d may be chosen to be independent of ξ, i, and j. In other words, $f_{ij}(x, u_i(x))$ is *locally onto*. This can be shown using an implicit function theorem type argument along with the fact that $\| (D_1 f_{ij})^{-1} \|$ is bounded.

It will be sufficient to show that

$$\| u_j(\xi') - u_j(\xi) - H_{ij} v_i(\xi)\,(\xi' - \xi) \| \le [\alpha\gamma(\beta a) + r(a)]\,\| \xi' - \xi \| \quad (3.109)$$

for all i, j, ξ, ξ' satisfying the above requirements with $\| \xi' - \xi \| \le a < d$. Then the result follows immediately by appealing to Lemma 3.3.8.

We begin with some preliminary estimates:

$$\begin{aligned}
\xi' - \xi &= f_{ij}(x', u_i(x')) - f_{ij}(x, u_i(x)) \\
&= C\,(x' - x) + E\,(u_i(x') - u_i(x)) \\
&\quad + \mathcal{O}\left(\| x' - x \|^2\right),
\end{aligned} \qquad (3.110)$$

$$\begin{aligned}
u_j(\xi') - u_j(\xi) &= g_{ij}(x', u_i(x')) - g_{ij}(x, u_i(x)) \\
&= A\,(x' - x) + B\,(u_i(x') - u_i(x)) \\
&\quad + \mathcal{O}\left(\| x - x' \|^2\right).
\end{aligned} \qquad (3.111)$$

From a previous estimate (Eq. (3.71)) we have

$$\| x' - x \| \le \frac{\| C^{-1} \|}{1 - 2\delta Q^2}\,\| \xi' - \xi \|. \qquad (3.112)$$

Also, rewriting (3.110) in the following form will be useful:

$$\begin{aligned}
\xi' - \xi &= (C + E v_i(x))\,(x - x') \\
&\quad + E\,(u_i(x') - u_i(x) - v_i(x)\,(x - x')) + \mathcal{O}\left(\| x - x' \|^2\right). (3.113)
\end{aligned}$$

We now begin the main estimates.

$$\| \, u_j(\xi') - u_j(\xi) - H_{ij} v_i(x) \, (\xi' - \xi) \, \|$$

$$= \; \| \, A \, (x - x') + B \, (u_i(x') - u_i(x)) + \mathcal{O} \left(\| \, x - x' \, \|^2 \right)$$

$$- [A + B v_i(x)] \, [C + E v_i(x)]^{-1} \, (\xi - \xi') \, \| \quad \text{from (3.111) and (3.92)}$$

$$= \; \| \, A \, (x - x') + B \, (u_i(x') - u_i(x)) + \mathcal{O} \left(\| \, x - x' \, \|^2 \right)$$

$$- [A + B v_i(x)] \, [C + E v_i(x)]^{-1} \, [(C + E v_i(x)) \, (x - x')$$

$$+ E \, (u_i(x') - u_i(x) - v_i(x) \, (x - x'))] \, \| \quad \text{from (3.113)}$$

$$= \; \| \, B \, (u_i(x') - u_i(x)) - B v_i(x) \, (x - x') - [A + B v_i(x)] \, [C + E v_i(x)]^{-1}$$

$$\times [E \, (u_i(x') - u_i(x) - v_i(x) \, (x - x'))] + \mathcal{O} \left(\| \, x - x' \, \|^2 \right) \, \|$$

$$= \; \| \, \left[B - [A + B v_i(x)] \, [C + E v_i(x)]^{-1} \, E \right]$$

$$\times [u_i(x') - u_i(x) - v_i(x) \, (x - x')] + \mathcal{O} \left(\| \, x - x' \, \|^2 \right) \, \|$$

$$\leq \; (\| \, B \, \| + \mathcal{O} \, (\delta)) \, \gamma \, (\| \, x' - x \, \|) \, \| \, x' - x \, \| + \mathcal{O} \left(\| \, x - x' \, \|^2 \right)$$

$$\leq \; \frac{[(\| \, B \, \| + \mathcal{O} \, (\delta)) \, \| \, C^{-1} \, \|]}{1 - 2\delta Q^2} \gamma \, (\| \, x' - x \, \|) \, \| \, \xi' - \xi \, \| + \mathcal{O} \left(\| \, x - x' \, \|^2 \right) ,$$

$$(3.114)$$

where the next to last inequality is obtained by using (3.108) and Proposition 3.3.5, and the last inequality is obtained by using (3.112).

Now the goal is to show that (3.114) can be put in the form of (3.109). Recall that

$$\| \, B \, \| \, \| \, C^{-1} \, \| < \frac{1}{2};$$

then if we define

$$\alpha \equiv \frac{(\| \, B \, \| + \mathcal{O} \, (\delta)) \, \| \, C^{-1} \, \|}{1 - 2\delta Q^2}, \qquad (3.115)$$

it follows that $\alpha < 1$ for η, δ sufficiently small. Next we define

$$\beta \equiv \frac{Q}{1 - 2\delta Q^2}. \tag{3.116}$$

Then from (3.112) we have

$$\| x - x' \| \leq \beta \| \xi' - \xi \| < \beta a. \tag{3.117}$$

Since $\gamma(\cdot)$ is nondecreasing, it follows that

$$\gamma \left(\| x - x' \| \right) \leq \gamma \left(\beta a \right). \tag{3.118}$$

From (3.117) it follows that

$$\mathcal{O} \left(\| x' - x \|^2 \right) < r(a) \| \xi' - \xi \| \tag{3.119}$$

for some function $r(a)$ where

$$r(a) \to 0 \qquad \text{as } a \to 0.$$

Combining (3.114), (3.115), (3.118), and (3.119), we have shown

$$\| u_j(\xi') - u_j(\xi) - H_{ij} v_i(\xi) \, (\xi' - \xi) \| \leq [\alpha \gamma \left(\beta a \right) + r(a)] \, \| \xi' - \xi \|$$

with the required hypotheses of Lemma 3.3.8. This proves the proposition.
□

Now we will proceed by induction to show that $\bar{M}^{\text{pert}}$ is C^r. We will next show that $\bar{M}^{\text{pert}}$ is C^2. At this step most of the difficulties associated with higher order derivatives will be clear. We will then suppose that $\bar{M}^{\text{pert}}$ is C^p, for $1 < p < r$, and show that this implies that $\bar{M}^{\text{pert}}$ is C^{p+1}.

The proof that $\bar{M}^{\text{pert}}$ is C^2 proceeds by the same steps that we used in showing that $\bar{M}^{\text{pert}}$ is C^1. We first derive a functional equation that the second derivative must satisfy and, using an iteration argument, show that this equation has a solution. Finally, we argue that this solution is, indeed, the second derivative.

We formally differentiate (3.96) and obtain the following equation which the second derivative must satisfy:

$$Dv_j(\xi) = \sum_{i=1}^{s} \left\{ D\phi_i \cdot [A + Bv_i(x)] \, [C + Ev_i(x)]^{-1} \right.$$

$$+ \phi_i \cdot D \left\{ [A + Bv_i(x)] \, [C + Ev_i(x)]^{-1} \right\} \right\} [C + Ev_i(x)]^{-1}$$

$$= \sum_{i=1}^{s} \left\{ D\phi_i \cdot [A + Bv_i(x)] \, [C + Ev_i(x)]^{-1} \right.$$

$$\phi_i \cdot \left\{ [DA + (DB)\, v_i(x) + BDv_i(x)]\, [C + Ev_i(x)]^{-1} \right.$$

$$- [A + Bv_i(x)]\, [C + Ev_i(x)]^{-1}\, [DC + (DE)\, v_i(x) + EDv_i(x)]$$

$$\left. [C + Ev_i(x)]^{-1} \right\} [C + Ev_i(x)]^{-1}, \tag{3.120}$$

where we have already taken the term $D\xi = D_1 f_{ij}(x, u_i(x)) + D_2 f_{ij}(x, u_i(x)) Du_i(x) = C + Ev_i(x)$ obtained via the chain rule from differentiating $v_j(\xi)$ with respect to x.

We then solve this equation iteratively; i.e., from (3.120), we pass to the following iteration scheme:

$$Dv_j^{n+1}(\xi) = \sum_{i=1}^{s} \left\{ D\phi_i \cdot [A + Bv_i^n(x)]\, [C + Ev_i^n(x)]^{-1} \right.$$

$$+\phi_i \cdot D\left\{ [A + Bv_i^n(x)]\, [C + Ev_i^n(x)]^{-1} \right\} \left. [C + Ev_i^n(x)]^{-1} \right.$$

$$= \sum_{i=1}^{s} \left\{ D\phi_i \cdot [A + Bv_i^n(x)]\, [C + Ev_i^n(x)]^{-1} \right.$$

$$\times \phi_i \cdot \left\{ [DA + (DB)\, v_i^n(x) + BDv_i^n(x)]\, [C + Ev_i^n(x)]^{-1} \right.$$

$$- [A + Bv_i^n(x)]\, [C + Ev_i^n(x)]^{-1}\, [DC + (DE)\, v_i^n(x) + EDv_i^n(x)]$$

$$\times [C + Ev_i^n(x)]^{-1} \right\} \left. [C + Ev_i^n(x)]^{-1}. \tag{3.121}$$

Technical Point: In (3.121) we have used the fact that

$$D\, [C + Ev_i^n(x)]^{-1} = - [C + Ev_i^n(x)]^{-1}\, D\, [C + Ev_i^n(x)]\, [C + Ev_i^n(x)]^{-1}.$$

This can be proven by noting that for an invertible map A, we have

$$D\left(AA^{-1}\right) = (DA)\, A^{-1} + AD\left(A^{-1}\right) = 0,$$

from which we conclude

$$D\left(A^{-1}\right) = -A^{-1}\, (DA)\, A^{-1}.$$

After rearranging the terms, (3.121) takes the form

$$Dv_j^{n+1}(\xi) = \sum_{i=1}^{s} \phi_i \cdot \left\{ BDv_i^n(x)\, [C + Ev_i^n(x)]^{-1} \right]$$

$$- [A + Bv_i^n(x)] [C + Ev_i^n(x)]^{-1}$$

$$\times E Dv_i^n(x) [C + Ev_i^n(x)]^{-1} \Big\} [C + Ev_i^n(x)]^{-1}$$

$$+ \sum_{i=1}^{s} \Big\{ D\phi_i \cdot [A + Bv_i^n(x)] [C + Ev_i^n(x)]^{-1}$$

$$+ \phi_i \cdot \Big\{ [DA + (DB) v_i^n(x)] [C + Ev_i^n(x)]^{-1}$$

$$- [A + Bv_i^n(x)] [C + Ev_i^n(x)]^{-1} [DC + (DE) v_i^n(x)]$$

$$\times [C + Ev_i^n(x)]^{-1} \Big\} [C + Ev_i^n(x)]^{-1} \Big\}. \tag{3.122}$$

The significance of this rearrangement is that the terms under the first summation sign in (3.122) depend on $Dv_i^n(x)$ and the terms under the second summation sign depend only on $v_i^n(x)$, which we previously showed is a Cauchy sequence. Thus, the second term has the general form $Z(v_i^n(x))$, where $Z(\cdot)$ is a continuous function of $v_i^n(x)$ and of x. Hence, it follows that

$$\lim_{n \to \infty} \| Z(v_i^n(x)) - Z(v_i^{n-1}(x)) \| = 0. \tag{3.123}$$

$D^2 u_i(x) \equiv Dv_i(x)$ is a bilinear map from $\mathbb{R}^{n-k}$ into $\mathbb{R}^k$. As such, it is an element of $L^2(\mathbb{R}^{n-k}, \mathbb{R}^k)$, which is naturally identified with $L(\mathbb{R}^{n-k}; L(\mathbb{R}^{n-k}, \mathbb{R}^k))$ (see Dieudonné [1960] or Abraham et al. [1988]), and this space has a natural operator norm, denoted $\| \cdot \|$, which we will use to study convergence. The second derivative of u is denoted by the following s-tuple:

$$D^2 u \equiv Dv = (Dv_1, \ldots, Dv_s),$$

and is an element of

$$[C^0(\mathcal{D}^3, L(\mathbb{R}^{n-k}; L(\mathbb{R}^{n-k}, \mathbb{R}^k)))]^s, \tag{3.124}$$

which is a complete metric space with the metric obtained from the following norm:

$$\| Dv \| \equiv \max_i \sup_{x \in \mathcal{D}^3} \| Dv_i(x) \|.$$

Now we want to show that $\{Dv^n\}$ is a Cauchy sequence. We begin by establishing several key estimates through a series of lemmas. Each of these lemmas is proved in the same way; namely, we break up the quantity to be estimated into terms which can be readily estimated using the following previously derived estimates:

$$\| A \|, \| B \|, \| C \|, \| C^{-1} \|, \| E \| \leq Q,$$
$$\| A \| < \eta,$$
$$\| B \| < \tfrac{1}{2},$$
$$\| B \| \| C^{-1} \| < \tfrac{1}{2}.$$

Using these estimates, along with Proposition 3.3.5, we easily obtain the following estimate (valid for δ sufficiently small) which will also prove useful:

$$1 - 2Q^2\delta \; < \| \left[\mathbb{1} + Ev_i^k C^{-1} \right]^{-1} \| < \; 1 + 2Q^2\delta.$$

Lemma 3.3.10 *For δ, η, and ϵ sufficiently small, $\{Dv^{n+1}\}$ is a bounded sequence.*

Proof: It suffices to show that $\| Dv_j^{n+1} \| \leq M, \; \forall n, j = 1, \ldots, s$. Using (3.122) and the estimates above, we easily obtain

$$
\begin{aligned}
\| Dv_j^{n+1} \| & \leq \; \left[\| B \| \| C^{-1} \|^2 + \mathcal{O}(\delta) \right] \| Dv_i^n \| + Z(v_i^n) \\
& \leq \; \left[\| B \| \| C^{-1} \|^2 + \mathcal{O}(\delta) \right] \| Dv^n \| + K, \quad (3.125)
\end{aligned}
$$

where Z is some continuous function of v_i^n and $K \equiv \sup_{i,n} \| Z(v_i^n) \|$. From this last expression, it follows that

$$
\| Dv_j^{n+1} \| \; \leq \; \left[\| B \| \| C^{-1} \|^2 + \mathcal{O}(\delta) \right]^n \| Dv^0 \|
$$

$$
+ K \sum_{i=0}^{n-1} \left[\| B \| \| C^{-1} \|^2 + \mathcal{O}(\delta) \right]^i. \quad (3.126)
$$

For η and δ sufficiently small, $\left[\| B \| \| C^{-1} \|^2 + \mathcal{O}(\delta) \right]$ is smaller than 1 (recall Eq. (3.52)). Hence, it follows from this last expression that $\| Dv_j^{n+1} \|$ is bounded. $\qquad \square$

Lemma 3.3.11 *For δ, η, and ϵ sufficiently small,*

$$
\| BDv_i^n(x) \left[C + Ev_i^n(x) \right]^{-1} \left[C + Ev_i^n(x) \right]^{-1}
$$

$$
- BDv_i^{n-1}(x) \left[C + Ev_i^{n-1}(x) \right]^{-1} \left[C + Ev_i^{n-1}(x) \right]^{-1} \|
$$

$$
\leq \left(\| B \| \| C^{-1} \|^2 + \mathcal{O}(\delta) \right) \| Dv_i^n(x) - Dv_i^{n-1}(x) \|
$$

$$
+ M_1 \| G_1(v_i^n) - G_1(v_i^{n-1}) \|, \quad (3.127)
$$

where $G_1(\cdot)$ is a continuous function of v_i^n and M_1 is some constant.

Proof: We have the following straightforward estimates:

$$\| BDv_i^n(x) \left[C + Ev_i^n(x)\right]^{-1} \left[C + Ev_i^n(x)\right]^{-1}$$

$$-BDv_i^{n-1}(x) \left[C + Ev_i^{n-1}(x)\right]^{-1} \left[C + Ev_i^{n-1}(x)\right]^{-1} \|$$

$$= \| \left(BDv_i^n(x) - BDv_i^{n-1}(x)\right) \left[C + Ev_i^{n-1}(x)\right]^{-1} \left[C + Ev_i^{n-1}(x)\right]^{-1}$$

$$+BDv_i^n(x) \left(\left[C + Ev_i^n(x)\right]^{-1} \left[C + Ev_i^n(x)\right]^{-1}\right.$$

$$\left. - \left[C + Ev_i^{n-1}(x)\right]^{-1} \left[C + Ev_i^{n-1}(x)\right]^{-1}\right) \|$$

$$\leq \| B \| \| C^{-1} \|^2 \left(1 + \mathcal{O}(\delta)\right) \| Dv_i^n - Dv_i^{n-1} \|$$

$$+ \| B \| \| Dv_i^n \| \| \left[C + Ev_i^n(x)\right]^{-1} \left[C + Ev_i^n(x)\right]^{-1}$$

$$- \left[C + Ev_i^{n-1}(x)\right]^{-1} \left[C + Ev_i^{n-1}(x)\right]^{-1} \|$$

$$\leq \left[\| B \| \| C^{-1} \|^2 + \mathcal{O}(\delta)\right] \| Dv_i^n - Dv_i^{n-1} \|$$

$$+M_1 \| G_1(v_i^n) - G_1(v_i^{n-1}) \|, \tag{3.128}$$

where for the constant M_1 and the function $G_1(\cdot)$ we make the appropriate identification on the next-to-last estimate. $\qquad\square$

Lemma 3.3.12 *For δ, η, and ϵ sufficiently small,*

$$\| \left[A + Bv_i^n(x)\right] \left[C + Ev_i^n(x)\right]^{-1} EDv_i^n(x) \left[C + Ev_i^n(x)\right]^{-1} \left[C + Ev_i^n(x)\right]^{-1}$$

$$- \left[A + Bv_i^{n-1}(x)\right] \left[C + Ev_i^{n-1}(x)\right]^{-1}$$

$$\times EDv_i^{n-1}(x) \left[C + Ev_i^{n-1}(x)\right]^{-1} \left[C + Ev_i^{n-1}(x)\right]^{-1} \|$$

$$\leq \mathcal{O}(\delta) \| Dv_i^n - Dv_i^{n-1} \| + M_2 \| G_2(v_i^n) - G_2(v_i^{n-1}) \|$$

$$\times M_3 \| G_3(v_i^n) - G_3(v_i^{n-1}) \| . \tag{3.129}$$

Proof:

$$\| \left[A + Bv_i^n(x)\right]\left[C + Ev_i^n(x)\right]^{-1} EDv_i^n(x)\left[C + Ev_i^n(x)\right]^{-1}$$

$$\times \left[C + Ev_i^n(x)\right]^{-1} - \left[A + Bv_i^{n-1}(x)\right]\left[C + Ev_i^{n-1}(x)\right]^{-1}$$

$$\times EDv_i^{n-1}(x)\left[C + Ev_i^{n-1}(x)\right]^{-1}\left[C + Ev_i^{n-1}(x)\right]^{-1} \|$$

$$= \| \left[A + Bv_i^n(x)\right]\left[C + Ev_i^n(x)\right]^{-1}\left(EDv_i^n(x) - EDv_i^{n-1}(x)\right)$$

$$\times \left[C + Ev_i^{n-1}(x)\right]\left[C + Ev_i^{n-1}(x)\right]^{-1}$$

$$+ \left[A + Bv_i^n(x)\right]\left[C + Ev_i^n(x)\right]^{-1} EDv_i^n(x)\left(\left[C + Ev_i^n(x)\right]^{-1}\right.$$

$$\times \left[C + Ev_i^n(x)\right]^{-1} - \left[C + Ev_i^{n-1}(x)\right]^{-1}\left[C + Ev_i^{n-1}(x)\right]^{-1}\right)$$

$$+ \left(\left[A + Bv_i^n(x)\right]\left[C + Ev_i^n(x)\right]^{-1} - \left[A + Bv_i^{n-1}(x)\right]\left[C + Ev_i^{n-1}(x)\right]^{-1}\right)$$

$$\times EDv_i^{n-1}(x)\left[C + Ev_i^{n-1}(x)\right]^{-1}\left[C + Ev_i^{n-1}(x)\right]^{-1} \|$$

$$\leq \| A + Bv_i^n(x) \| \,\| \left[C + Ev_i^n(x)\right]^{-1} \| \,\| E \|$$

$$\times \| Dv_i^n - Dv_i^{n-1} \| \,\| \left[C + Ev_i^{n-1}(x)\right]^{-1} \| \,\| \left[C + Ev_i^{n-1}(x)\right]^{-1} \|$$

$$+ \| A + Bv_i^n(x) \| \,\| \left[C + Ev_i^n(x)\right]^{-1} \| \,\| E \| \,\| Dv_i^n \|$$

$$\times \| \left[C + Ev_i^n(x)\right]^{-1}\left[C + Ev_i^n(x)\right]^{-1}$$

$$- \left[C + Ev_i^{n-1}(x)\right]^{-1}\left[C + Ev_i^{n-1}(x)\right]^{-1} \|$$

$$+ \| \left[A + Bv_i^n(x)\right]\left[C + Ev_i^n(x)\right]^{-1}$$

$$- \left[A + Bv_i^{n-1}(x)\right]\left[C + Ev_i^{n-1}(x)\right]^{-1} \| \,\| E \|$$

$$\times \| Dv_i^{n-1} \| \,\| \left[C + Ev_i^{n-1}(x)\right]^{-1} \| \,\| \left[C + Ev_i^{n-1}(x)\right]^{-1} \|$$

$$\leq \mathcal{O}(\delta) \parallel Dv_i^n - Dv_i^{n-1} \parallel + M_2 \parallel G_2(v_i^n) - G_2(v_i^n) \parallel$$

$$\times M_3 \parallel G_3(v_i^n) - G_3(v_i^n) \parallel, \tag{3.130}$$

where for the constants M_2 and M_3 and the functions $G_2(\cdot)$ and $G_3(\cdot)$, we make the appropriate identification on the next-to-last estimate.   □

Proposition 3.3.13 *For δ, η, and ϵ sufficiently small,*

$$\parallel Dv_j^{n+1} - Dv_j^n \parallel \leq [\parallel B \parallel \parallel C^{-1} \parallel^2 + \mathcal{O}(\delta)] \parallel Dv_i^n - Dv_i^{n-1} \parallel$$

$$+ M_1 \parallel G_1(v_i^n) - G_1(v_i^{n-1}) \parallel + M_2 \parallel G_2(v_i^n) - G_2(v_i^{n-1}) \parallel$$

$$+ M_3 \parallel G_3(v_i^n) - G_3(v_i^{n-1}) \parallel + \parallel Z(v_i^n) - Z(v_i^{n-1}) \parallel \tag{3.131}$$

□

Proof: This follows immediately from (3.122) and Lemmas 3.3.10, 3.3.11, and 3.3.12.   □

Finally, we can prove the main result concerning the second derivative.

Proposition 3.3.14 *For δ, η, and ϵ sufficiently small, $\{Dv^n\}$ convergences to an element Dv, which is the second derivative of u. Moreover, the components of Dv satisfy (3.122).*

Proof: We know that $\{v^n\}$ is a Cauchy sequence that converges to v, and that v is the derivative of u. Moreover, each v^n is differentiable. By assumption, we have $\parallel B \parallel \parallel C^{-1} \parallel^2 < \frac{1}{2}$ (recall (3.52)); therefore, for η and δ sufficiently small, $\parallel B \parallel \parallel C^{-1} \parallel^2 + \mathcal{O}(\delta) < \frac{3}{4}$. Hence, it follows from Proposition 3.3.13 that $\{Dv^n\}$ is a Cauchy sequence, so it converges to some Dv. We would like to conclude that $Dv = D^2u$. This follows immediately from the fact that the convergence of $\{Dv^n\}$ is uniform and each v^n is differentiable (see Rudin [1964], Theorem 7.17).

The fact that the components of v satisfy (3.122) follows from construction.   □

Now we consider the case of the pth derivative of u, $1 \leq p \leq r$. $D^p u_i(x)$ is a p-linear map from $\mathbb{R}^{n-k}$ into $\mathbb{R}^k$. So, as earlier, $D^p u$ is represented as an s-tuple $D^p u = (D^p u_1, \ldots, D^p u_s)$ with

$$D^p u \in [C^0(\mathcal{D}^3, \text{L}^p(\mathbb{R}^{n-k}, \mathbb{R}^k))]^s.$$

We use the natural operator norm, denoted $\| \cdot \|$, on the space $L^p\left(\mathbb{R}^{n-k}, \mathbb{R}^k\right)$ and for $w = (w_1, \ldots, w_s) \in [C^0(\mathcal{D}^3, L^p(\mathbb{R}^{n-k}, \mathbb{R}^k))]^s$ we define $\| w \| = \max_i \sup_{x \in \mathcal{D}^3} \| w_i(x) \|$.

Recall the functional equation satisfied by the components of the second derivative of u (Eq. (3.122)), which we rewrite

$$Dv_j(\xi) = \sum_{i=1}^s \phi_i \cdot \left\{ BDv_i(x) \left[C + Ev_i(x)\right]^{-1} \right.$$

$$- \left[A + Bv_i(x)\right] \left[C + Ev_i(x)\right]^{-1}$$

$$\times EDv_i(x) \left[C + Ev_i(x)\right]^{-1} \right\} \left[C + Ev_i(x)\right]^{-1}$$

$$+ \sum_{i=1}^s \left\{ D\phi_i \cdot \left[A + Bv_i(x)\right] \left[C + Ev_i(x)\right]^{-1} \right.$$

$$+ \phi_i \cdot \left\{ \left[DA + (DB)\, v_i(x)\right] \left[C + Ev_i(x)\right]^{-1} \right.$$

$$- \left[A + Bv_i(x)\right] \left[C + Ev_i(x)\right]^{-1} \left[DC + (DE)\, v_i(x)\right]$$

$$\times \left[C + Ev_i(x)\right]^{-1} \right\} \left[C + Ev_i(x)\right]^{-1} \right\}. \qquad (3.132)$$

Let us differentiate both the left- and right-hand sides of (3.132) $p-2$ times, and set up the usual iteration scheme to obtain

$$D^{p-1} v_j^{n+1}(\xi) = \sum_{i=1}^s \phi_i \cdot \left\{ B D^{p-1} v_i^n(x) \left[C + Ev_i^n(x)\right]^{-1} \right.$$

$$- \left[A + Bv_i^n(x)\right] \left[C + Ev_i^n(x)\right]^{-1}$$

$$\times E D^{p-1} v_i^n(x) \left[C + Ev_i^n(x)\right]^{-1} \right\}$$

$$\times \underbrace{\left[C + Ev_i^n(x)\right]^{-1} \cdots \left[C + Ev_i^n(x)\right]^{-1}}_{p\text{-1 terms}}$$

$$+ F(v_i^n, Dv_i^n, D^2 v_i^n, \ldots, D^{p-2} v_i^n), \qquad (3.133)$$

where $F(\cdot)$ is a continuous function of $v_i^n, Dv_i^n, D^2 v_i^n, \ldots, D^{p-2} v_i^n$ (provided, of course, that the pth derivative of u exists). The additional $p-2$ terms on the right-hand side of (3.132) of the form $\left[C + Ev_i^n(x)\right]^{-1}$

arise as a result of the use of the chain rule in differentiating the left-hand side of (3.132). From the form of (3.133), and using the estimate $\| [C + Ev_i^n(x)]^{-1} \| \leq \| C^{-1} \| + \mathcal{O}(\delta)$, it is clear that lemmas analogous to 3.3.10, 3.3.11, and 3.3.12 can be proven along the exact same lines. From this, we easily obtain an estimate of the form

$$\| D^{p-1}v_j^{n+1} - D^{p-1}v_j^n \| \leq \left[\| B \| \, \| C^{-1} \|^p + \mathcal{O}(\eta + \delta) \right]$$

$$\times \| D^{p-1}v_i^n - D^{p-1}v_i^{n-1} \| + \| F(v_i^n, Dv_i^n, D^2v_i^n, \ldots, D^{p-2}v_i^n)$$

$$-F(v_i^{n-1}, Dv_i^{n-1}, D^2v_i^{n-1}, \ldots, D^{p-2}v_i^{n-1}) \| \, . \tag{3.134}$$

Thus, we argue inductively. Assuming u is C^{p-1}, these estimates hold and since, by assumption, $\| B \| \, \| C^{-1} \|^p < \frac{1}{2}$, $1 \leq p \leq r$ (recall (3.52)) the argument given in Proposition 3.3.14 holds. This completes the proof of Theorem 3.3.1.

4

The Unstable Manifold of an Overflowing Invariant Manifold

It is reasonable to consider the existence of the unstable manifold (but *not* the stable manifold) of an overflowing invariant manifold. In this chapter, under appropriate hypotheses, we will construct the unstable manifold of the overflowing invariant manifold $\bar{M}$. Intuitively, we think of the unstable manifold of an invariant set as the set of points in phase space that approach the invariant set as $t \to -\infty$. This notion requires careful interpretation since $\bar{M}$ has a boundary, and trajectories starting on $\bar{M}$ may leave $\bar{M}$ in forward time by crossing ∂M (this is the reason that it does not make sense to consider a stable manifold). We also want to emphasize that initially we will *not* be dealing with a perturbation problem; we will be concerned with constructing the unstable manifold of $\bar{M}$. Afterward, we will show that this unstable manifold also satisfies the hypotheses of the persistence theorem for overflowing invariant manifolds. Hence, $\bar{M}^{\mathrm{pert}}$ will also have an unstable manifold under appropriate hypotheses. We begin developing the setting in much the same way as earlier.

4.1 Generalized Lyapunov-Type Numbers

Suppose we have the continuous splitting

$$T\mathbb{R}^n|_{M_2} = TM_2 \oplus N^s \oplus N^u$$

and the associated projections

$$\Pi^u \quad : \quad T\mathbb{R}^n|_{M_2} \to N^u, \tag{4.1}$$

$$\Pi^s \quad : \quad T\mathbb{R}^n|_{M_2} \to N^s. \tag{4.2}$$

For notational purposes we define $N \equiv N^s \oplus N^u$. We assume that the subbundles $TM_2 \oplus N^u$ and $TM_2 \oplus N^s$ are each invariant under $D\phi_t$ for all $t < 0$ (i.e., overflowing invariant). Growth rates of vectors in these subbundles under the linearized dynamics are characterized by generalized

Lyapunov-type numbers as before. For a point $p \in M_2$, we consider the following *nonzero* vectors:

$$
\begin{aligned}
u_0 &\in N_p^u, \\
w_0 &\in N_p^s, \\
v_0 &\in T_p M_2,
\end{aligned}
$$

and

$$
\begin{aligned}
u_{-t} &= \Pi^u D\phi_{-t}(p) u_0, \\
w_{-t} &= \Pi^s D\phi_{-t}(p) w_0, \\
v_{-t} &= D\phi_{-t}(p) v_0.
\end{aligned}
$$

We define the following generalized Lyapunov-type numbers:

$$
\lambda^u(p) = \inf \left\{ a : \left(\frac{\| u_{-t} \|}{\| u_0 \|} \right) / a^t \to 0 \quad \text{as} \quad t \uparrow \infty, \quad \forall u_0 \in N_p^u \right\}, \quad (4.3)
$$

$$
\nu^s(p) = \inf \left\{ a : \left(\frac{\| w_0 \|}{\| w_{-t} \|} \right) / a^t \to 0 \quad \text{as} \quad t \uparrow \infty, \quad \forall w_0 \in N_p^s \right\}. \quad (4.4)
$$

If $\nu^s(p) < 1$, we then define

$$
\begin{aligned}
\sigma^s(p) \;=\; &\inf \left\{ b : \left(\| w_0 \|^b / \| v_0 \| \right) / \left(\| w_{-t} \|^b / \| v_{-t} \| \right) \to 0 \right. \\
&\left. \text{as} \quad t \uparrow \infty, \quad \forall v_0 \in T_p M_2, \; w_0 \in N_p^s \right\}. \quad (4.5)
\end{aligned}
$$

Following the same calculations as in Lemma 3.1.1, we can show that these expressions are equal to the following:

$$
\lambda^u(p) \;=\; \overline{\lim_{t \to \infty}} \, \| \Pi^u D\phi_{-t}(p) |_{N_p^u} \|^{\frac{1}{t}}, \quad (4.6)
$$

$$
\nu^s(p) \;=\; \overline{\lim_{t \to \infty}} \, \| \Pi^s D\phi_t \left(\phi_{-t}(p) \right) |_{N_p^s} \|^{\frac{1}{t}}, \quad (4.7)
$$

$$
\sigma^s(p) \;=\; \overline{\lim_{t \to \infty}} \, \frac{\log \| D\phi_{-t} |_{M_2}(p) \|}{-\log \| \Pi^s D\phi_t \left(\phi_{-t}(p) \right) |_{N_p^s} \|}. \quad (4.8)
$$

Lemma 3.1.2 (constant on orbits) and Proposition 3.1.3 (independent of metric) also apply to these generalized Lyapunov-type numbers. The proofs involve identical calculations.

We say that the splitting is *hyperbolic* if

$$
\lambda^u(p) < 1, \qquad \nu^s(p) < 1, \qquad \forall p \in M. \quad (4.9)
$$

The geometrical interpretation of these generalized Lyapunov-type numbers should be clear from the earlier discussion. Moreover, we have the following *uniformity lemma* due to Fenichel [1971].

Lemma 4.1.1 (Uniformity Lemma for Unstable Manifolds)

1. *Suppose*

$$\| \Pi^s D\phi_t \left(\phi_{-t}(p) \right) |_{N_p^s} \| / a^t \to 0 \quad as \quad t \uparrow \infty, \quad \forall p \in \bar{M}_1.$$

   Then there are constants $\hat{a} < a$ and C such that

$$\| \Pi^s D\phi_t \left(\phi_{-t}(p) \right) |_{N_p^s} \| < C\hat{a}^t, \quad \forall p \in \bar{M}_1 \quad and \quad t \geq 0.$$

2. *Suppose*

$$\| \Pi^u D\phi_{-t}(p) |_{N_p^u} \| / a^t \to 0 \quad as \quad t \uparrow \infty, \quad \forall p \in \bar{M}_1.$$

   Then there are constants $\hat{a} < a$ and C such that

$$\| \Pi^u D\phi_{-t}(p) |_{N_p^u} \| < C\hat{a}^t, \quad \forall p \in \bar{M}_1 \quad and \quad t \geq 0.$$

3. *Under the hypotheses of (1), suppose also that $a \leq 1$ and*

$$\| D\phi_{-t}|_{M_2}(p) \| \, \| \Pi^s D\phi_t \left(\phi_{-t}(p) \right) |_{N_p^s} \|^b \to 0 \quad as \quad t \uparrow \infty, \quad \forall p \in \bar{M}_1.$$

   Then there are constants $\hat{b} < b$ and C such that

$$\| D\phi_{-t}|_{M_2}(p) \| \, \| \Pi^s D\phi_t \left(\phi_{-t}(p) \right) |_{N_p^s} \|^{\hat{b}} < C, \quad \forall p \in \bar{M}_1 \quad and \quad t \geq 0.$$

4. *If $\nu^s(p) < a \leq 1$, $\lambda^u(p) < a \leq 1$, and $\sigma^s(p) < b \; \forall p \in \bar{M}_1$, then*

$$\| \Pi^s D\phi_t \left(\phi_{-t}(p) \right) |_{N_p^s} \| \to 0 \quad as \quad t \uparrow \infty,$$

$$\| \Pi^u D\phi_{-t}(p) |_{N_p^u} \| \to 0 \quad as \quad t \uparrow \infty,$$

   and

$$\| D\phi_{-t}|_{M_2}(p) \| \, \| \Pi^s D\phi_t \left(\phi_{-t}(p) \right) |_{N_p^s} \|^b \to 0 \quad as \quad t \uparrow \infty$$

   uniformly $\forall p \in \bar{M}_1$.

5. $\nu^s(\cdot)$, $\lambda^u(\cdot)$, and $\sigma^s(\cdot)$ attain their suprema on M.

Proof: The proof proceeds exactly as in the proof of the uniformity lemma given earlier.   □

4.2 Local Coordinates Near M

Next we need to construct local coordinates near M_2, describe the flow in these coordinates, and set up the graph transform. Much of the construction will be similar to that given earlier, so we will not give all the details.

We begin by constructing a neighborhood of M. The dimension of N_p^s will be s and the dimension of N_p^u will be u. Hence, the dimension of M will be $n - (s+u)$. Following Proposition 3.2.3, we perturb the bundles N^s and N^u to C^r bundles N'^s and N'^u, respectively.

> **Perturbation of the Subbundles N^s and N^u:** Proposition 3.2.3 showed how the C^{r-1} normal bundle N could be perturbed to a C^r *transversal bundle* N'. Exactly the same argument applies to $N \equiv N^s \oplus N^u$. Following the same notation as in Proposition 3.2.3, we define maps
>
> $$\begin{aligned} \Phi_u &: \quad \bar{M}_2 \to G(u, n), \\ \Phi_s &: \quad \bar{M}_2 \to G(s, n). \end{aligned}$$
>
> Following exactly the same arguments, one can construct C^r maps $\tilde{\Phi}_u$ and $\tilde{\Phi}_s$ that are arbitrarily C^0 close to Φ_u and Φ_s, respectively, on $\bar{M}_1$. One can then conclude that there are C^r subbundles N'^u and N'^s that are C^0 ϵ-close to N^u and N^s, respectively, with $N' \equiv N'^u \oplus N'^s$. Moreover, N'^u and N'^s have C^r orthonormal bases.

We then define

$$\begin{aligned} N'^s_\epsilon &\equiv \quad \{(p, v^s) \in N'^s : \| v^s \| \le \epsilon \}, \\ N'^u_\epsilon &\equiv \quad \{(p, v^s) \in N'^u : \| v^u \| \le \epsilon \}, \end{aligned}$$

and the map

$$\begin{aligned} h &: \quad N'^s_\epsilon \oplus N'^u_\epsilon \to \mathbb{R}^n, \\ &(p, v^s, v^u) \mapsto p + v^s + v^u. \end{aligned} \tag{4.10}$$

From Proposition 3.2.4 it follows that there is an $\epsilon_0 > 0$ such that for all $0 < \epsilon \le \epsilon_0$ the map h is a C^r map, mapping $N'^s_\epsilon \oplus N'^u_\epsilon |_{\cup_{i=1}^s U_i^5}$ onto a neighborhood of $\cup_{i=1}^s \bar{U}_i^4$.

For each i we choose C^r orthonormal bases for N'^s_ϵ and N'^u_ϵ (recall Corollary 3.2.2) and we define the maps

$$\tau_i^s \quad : \quad N'^s_\epsilon|_{U_i^6} \to \mathbb{R}^s,$$

$$(p, v^s) \to \left\{ \begin{array}{l} \text{vector of coordinates of } v^s \text{ with respect to the} \\ \text{chosen basis of } N'^s_\epsilon|_p. \end{array} \right\}$$

and

$$\tau_i^u \quad : \quad N'^u_\epsilon|_{U_i^6} \to \mathbb{R}^u,$$

$$(p, v^u) \to \left\{ \begin{array}{l} \text{vector of coordinates of } v^s \text{ with respect to the} \\ \text{chosen basis of } N'^u_\epsilon|_p. \end{array} \right\}$$

We use these to define the map

$$\sigma_i \times \tau_i^s \times \tau_i^u : N'_\epsilon|_{U_i^6} \to \mathbb{R}^{n-(s+u)} \times \mathbb{R}^s \times \mathbb{R}^u,$$
$$(\sigma_i \times \tau_i^s \times \tau_i^u)(p, v^s, v^u) = (\sigma_i(p), \tau_i^s(p, v^s), \tau_i^u(p, v^u)) \tag{4.11}$$
$$\equiv (x, y, z),$$

which provide local coordinates near a neighborhood of $\bar{U}_i^4$ in $\mathbb{R}^n$ (recall Proposition 3.2.5).

In these local coordinates the time-T flow map is given by

$$(x, y, z) \to \left(f_{ij}^0(x, y, z), g_{ij}^0(x, y, z), h_{ij}^0(x, y, z) \right), \tag{4.12}$$

where

$$\begin{aligned} f_{ij}^0(x, y, z) &= \sigma_j \circ \phi_T \circ (\sigma_i \times \tau_i^s \times \tau_i^u)^{-1}(x, y, z), \\ g_{ij}^0(x, y, z) &= \tau_j^s \circ \phi_T \circ (\sigma_i \times \tau_i^s \times \tau_i^u)^{-1}(x, y, z), \\ h_{ij}^0(x, y, z) &= \tau_j^u \circ \phi_T \circ (\sigma_i \times \tau_i^s \times \tau_i^u)^{-1}(x, y, z), \end{aligned} \tag{4.13}$$

which are defined for

$$(x, y, z) \in (\sigma_i \times \tau_i^s \times \tau_i^u) N'_\epsilon|_{\overline{U_i^4 \cap \phi_{-T}\left(\overline{U_j^4}\right)}}.$$

Note the explicit use of the superscript 0 in our notation for the time-T map. As in our development of the background and notation for the proof of the persistence theorem for overflowing invariant manifolds, this superscript 0 refers to the unperturbed dynamics. At this stage we are not considering any perturbation to our vector field.

4.3   The GR-I Estimates

We now give a series of estimates related to various components of the time-T map and its partial derivatives that will be important for the success of our contraction mapping arguments. The estimates are grouped according to the various geometrical and dynamical features from which they arise.

Estimates Derived from Growth Rates

Henceforth, we will assume that

$$\boxed{\lambda^u(p) < 1, \ \nu^s(p) < 1, \text{ and } \sigma^s(p) < \tfrac{1}{r} \text{ for every } p \in \bar{M}_1.}$$

It follows from this assumption and the uniformity lemma that there is a $T > 1$ such that

$$\| \ \Pi'^s D\phi_T\left(\phi_{-T}(p)\right)|_{N_p^s} \ \| < \tfrac{1}{8},$$

$$\| \ \Pi'^u D\phi_{-T}(p)|_{N_p^u} \ \| < \tfrac{1}{4},$$

$$c^{2k} \ \| \ D\phi_{-T}|_{M_2}(p) \ \|^k \| \ \Pi'^s D\phi_T\left(\phi_{-T}(p)\right)|_{N_p^s} \ \| < \tfrac{1}{8}, \qquad 0 \le k \le r.$$
$$(4.14)$$

We fix this T for the rest of our constructions and arguments. Next, we translate these estimates into the local coordinates using our local expressions for the time-T map given above:

$$\| \ D_2 g^0_{ij}(x,0,0) \ \| < \tfrac{1}{4},$$

$$\| \left(D_3 h^0_{ij}(x,0,0)\right)^{-1} \ \| < \tfrac{1}{8}, \qquad\qquad\qquad (4.15)$$

$$\| \left(D_1 f^0_{ij}(x,0,0)\right)^{-1} \ \|^k \| \ D_2 g^0_{ij}(x,0,0) \ \| < \tfrac{1}{8}, \qquad 1 \le k \le r.$$

For ϵ sufficiently small (which in local coordinates means y and z sufficiently small) we have by continuity

$$\| \ D_2 g^0_{ij}(x,y,z) \ \| < \tfrac{1}{2},$$

$$\| \left(D_3 h^0_{ij}(x,y,z)\right)^{-1} \ \| < \tfrac{1}{4}, \qquad\qquad\qquad (4.16)$$

$$\| \left(D_1 f^0_{ij}(x,y,z)\right)^{-1} \ \|^k \| \ D_2 g^0_{ij}(x,y,z) \ \| < \tfrac{1}{4}, \qquad 1 \le k \le r.$$

Estimates Derived from the Overflowing Invariance of M Under the Nonlinear Dynamics

Overflowing invariance of M is manifested in the local coordinate expression of the time-T map as

$$g_{ij}^0(x, 0, 0) = 0,$$
$$h_{ij}^0(x, 0, 0) = 0. \tag{4.17}$$

From these expressions it immediately follows that

$$D_1 g_{ij}^0(x, 0, 0) = 0,$$
$$D_1 h_{ij}^0(x, 0, 0) = 0. \tag{4.18}$$

It follows then, by continuity, that for given $\eta > 0$ there exists $\epsilon > 0$ such that if $\| (\tau_i^s)^{-1}(y) \| < \epsilon$ and $\| (\tau_i^u)^{-1}(z) \| < \epsilon$, then we have

$$\| g_{ij}^0(x, y, z) \| < \eta,$$
$$\| h_{ij}^0(x, y, z) \| < \eta,$$
$$\| D_1 g_{ij}^0(x, y, z) \| < \eta,$$
$$\| D_1 h_{ij}^0(x, y, z) \| < \eta. \tag{4.19}$$

Estimates Derived from the Overflowing Invariance of $TM_2 \oplus N^s$ and $TM_2 \oplus N^u$ Under the Linearized Dynamics

First note that in the title of this subsection we have used N^s and N^u rather than N'^s and N'^u. This technical distinction is important as $TM_2 \oplus N^s$ and $TM_2 \oplus N^u$ are overflowing invariant under the linearized dynamics; $TM_2 \oplus N'^s$ and $TM_2 \oplus N'^u$ need not be.

The derivative of the time-T map in the local coordinates (which were defined using N'^s and N'^u) is

$$\begin{pmatrix} D_1 f_{ij}^0 & D_2 f_{ij}^0 & D_3 f_{ij}^0 \\ D_1 g_{ij}^0 & D_2 g_{ij}^0 & D_3 g_{ij}^0 \\ D_1 h_{ij}^0 & D_2 h_{ij}^0 & D_3 h_{ij}^0 \end{pmatrix}, \tag{4.20}$$

where we have (for the moment) deliberately omitted displaying the points at which the entries of this matrix are evaluated. In local coordinates the fibers of N'^s are denoted by $(x, y, 0)$ and the fibers of N'^u are denoted by $(x, 0, z)$.

Now suppose that we instead have used N^s and N^u to construct local coordiantes for the representation of the time-T map. Then invariance of N^s under the linearized dynamics implies that

$$D_1 h_{ij}^0(x, y, 0) = 0,$$

$$D_2 h_{ij}^0(x, y, 0) = 0,$$
(4.21)

and invariance of N^u under the linearized dynamics implies that

$$D_1 g_{ij}^0(x, 0, z) = 0,$$

$$D_3 g_{ij}^0(x, 0, z) = 0.$$
(4.22)

Now the bundle N'^s (resp. N'^u) can be chosen to be arbitrarily close to N^s (resp. N^u); hence, for a given $\eta > 0$, and ϵ sufficiently small, we have

$$\| D_1 h_{ij}^0(x, y, z) \| < \eta,$$

$$\| D_2 h_{ij}^0(x, y, z) \| < \eta,$$

$$\| D_1 g_{ij}^0(x, y, z) \| < \eta,$$
(4.23)

$$\| D_3 g_{ij}^0(x, y, z) \| < \eta.$$

A Dilational Estimate

The idea for the following estimate comes from Fenichel [1974], p. 1133. We want to argue that $\| D_3 f_{ij}^0 \|$ can be made small by an appropriate rescaling of the z coordinate. Consider the following dilation of the coordinates:

$$(x, y, z) \mapsto (x, y, \kappa z) = (x, y, \bar{z}),$$
(4.24)

where κ is a positive real number. First we want to consider how this dilation will affect the estimates (4.16), (4.19), and (4.23) given above. It is easy to see that it has no effect on the estimates (4.19) derived from overflowing invariance of M under the nonlinear dynamics. The only estimate in (4.23) affected by the dilation of coordinates is $\| D_3 g_{ij}^0(x, y, \bar{z}) \| < \eta$, which becomes $\| D_3 g_{ij}^0(x, y, \bar{z}) \| < \kappa \eta$ in the dilated coordinates.

However, the estimates derived from the growth rates, (4.16), require more careful consideration. For these three quantities, one need only worry about terms involving the partial derivative with respect to z (D_3) in particular, the quantity $\|(D_3 h_{ij}^0(x, 0, 0))^{-1}\| < \frac{1}{8}$. In the dilated coordinates this estimate becomes $\|(D_3 h_{ij}^0(x, 0, 0))^{-1}\| < \frac{\kappa}{8}$, which is smaller than $\frac{1}{8}$ for $\kappa < 1$.

Now we turn to the quantity of interest. In the dilated coordinates it becomes $\kappa \parallel D_3 f_{ij}^0 (x,0,0) \parallel$, which clearly can be made arbitrarily small, say less than η, by taking κ sufficiently small. Henceforth, we will assume that this holds. Moreover, by continuity, we may also further assume that for x and y sufficiently small we have

$$\parallel D_3 f_{ij}^0 (x, y, \bar{z}) \parallel < \eta. \tag{4.25}$$

We henceforth assume that we are working in these dilated coordinates and drop the "bar" from our notation on z.

In the following arguments we will have reason to refer to the estimates (4.16), (4.19), (4.23), and (4.25) repeatedly. In such instances we will collectively refer to these estimates as the **GR-I estimates** *(for "growth rate" and "invariance").*

General Bounds

We assume that the norms of the first and second partial derivatives, and the inverses (when appropriately defined), of f_{ij}^0, g_{ij}^0, and h_{ij}^0 are bounded by some constant, say Q. This follows from the fact that these functions are C^r and we are restricting ourselves to a compact subset of $\mathbb{R}^n$.

4.4   The Space of Sections of $N'_\epsilon|_{\cup_{i=1}^s U_i^3}$ over $N'^u_\epsilon|_{\cup_{i=1}^s U_i^3}$

The unstable manifold of M will be constructed as the graph of a section u of $N'_\epsilon|_{\cup_{i=1}^s U_i^3}$ over $N'^u_\epsilon|_{\cup_{i=1}^s U_i^3}$. However, first we should describe how we view N'_ϵ as a vector bundle over N'^u_ϵ.

A point in $N'_\epsilon \equiv N'^u_\epsilon \oplus N'^s_\epsilon$ is denoted by (p, v^u, v^s), where $p \in M_2$. Thus, viewing N'^u_ϵ as the base space of the vector bundle N'_ϵ, a point in N'_ϵ is denoted by (p, v^u, v^s), where (p, v^u) is regarded as the basepoint. A section of N'_ϵ over N'^u_ϵ is a map

$$\begin{aligned} u : N'^u_\epsilon &\rightarrow N'_\epsilon, \\ (p, v^u) &\mapsto (p, v^u, v^s), \end{aligned}$$

and let S denote the space of sections of $N'_\epsilon|_{\cup_{i=1}^s U_i^3}$ over $N'^u_\epsilon|_{\cup_{i=1}^s U_i^3}$.
Local coordinate representations of a section u are given by

$$u_i(x, z) = \tau_i^s \circ \Pi'^s \circ u \circ (\sigma_i \times \tau_i^u)^{-1} (x, z),$$

where

$$u_i : \mathcal{D}^3 \times \mathcal{D}^\epsilon \to \mathbb{R}^s$$

and $\mathcal{D}^\epsilon \equiv \{z \in \mathbb{R}^u \mid \parallel z \parallel \leq \epsilon\}$. Since the notation $\mathcal{D}^\epsilon$ arises here for the first time, it is perhaps worthwhile to discuss its origins. According to the above formulas, the domain of the local coordinate representation of u is $\mathcal{D}^3 \times \mathcal{D}^\epsilon$. This follows from the fact that we are using orthornormal bases to locally describe points in N'^s_ϵ and N'^u_ϵ, from which it follows that $\parallel \tau^u(p, v^u) \parallel = \parallel v^u \parallel = \parallel z \parallel$. Using this, we have

$$(\sigma_i \times \tau_i^u)\left(N'_\epsilon|_{U_i^j}\right) \equiv \sigma_i\left(U_i^j\right) \times \tau_i^u\left(N'_\epsilon|_{U_i^j}\right)$$
$$\equiv \mathcal{D}^j \times \mathcal{D}^\epsilon,$$

which is the domain for the local coordinate representation of u. Thus, in coordinates, the section u is represented by the s maps U_i, $i = 1, \ldots, s$ each defined on $\mathcal{D}^3 \times \mathcal{D}^\epsilon$. We define

$$\text{Lip}\, u = \max_i \sup_{\substack{(x,z),(x',z') \in \mathcal{D}^3 \times \mathcal{D}^\epsilon \\ (x,z) \neq (x',z')}} \frac{\parallel u_i(x, z) - u_i(x', z') \parallel}{\parallel x - x' \parallel + \parallel z - z' \parallel}, \qquad (4.26)$$

if this exists, and denote

$$S_\delta = \{u \in S \mid \text{Lip}\, u \leq \delta\}. \qquad (4.27)$$

Note that we have

$$\parallel u_i(x, z) - u_i(x', z') \parallel \leq \delta\, (\parallel x - x' \parallel + \parallel z - z' \parallel). \qquad (4.28)$$

It is straightforward to show that S_δ is a complete metric space with the metric derived from the C^0 norm.

4.5   The Unstable Manifold Theorem

Before stating the unstable manifold theorem we need to address a technical point.

> **Technical Point:** If one thinks of the situation of the stable and unstable manifolds of a hyperbolic fixed point of a vector field, the unstable manifold of the fixed point is tangent to the unstable subspace of the linearized vector field, at the fixed point. Similarly, in this general setting, the unstable manifold of $\bar{M}$ will be tangent to some "linear object" at $\bar{M}$—but what is this "linear object"? The obvious candidate is N'^u_ϵ since trajectories with initial conditions in this bundle have the appropriate asymptotic behavior as $t \to -\infty$ and we will construct the unstable manifold as a graph over N'^u_ϵ.

However, there is a problem with this choice as the bundle N'^u_ϵ is not naturally contained in $\mathbb{R}^n$. This issue can be easily handled by embedding the bundle in $\mathbb{R}^n$ in a natural way.

We define the map

$$
\begin{aligned}
h_u : N'^u_\epsilon &\rightarrow \mathbb{R}^n, \\
(p, v^u) &\mapsto p + v^u.
\end{aligned}
$$

Then $h_u\left(N'^u_\epsilon\right) \subset \mathbb{R}^n$ and we can consider the issue of the tangency of the unstable manifold of $\bar{M}$ and $h_u\left(N'^u_\epsilon\right)$.

We now state the unstable manifold theorem.

Theorem 4.5.1 (Fenichel, 1971) *Suppose $\dot{x} = f(x)$ is a C^r vector field on $\mathbb{R}^n$, $r \geq 1$. Let $\bar{M} \equiv M \cup \partial M$ be a C^r, compact connected manifold with boundary overflowing invariant under the vector field $f(x)$. Suppose $\nu^s(p) < 1$, $\lambda^u(p) < 1$, and $\sigma^s(p) < \frac{1}{r}$ for all $p \in M$. Then there exists a C^r overflowing invariant manifold $W^u(\bar{M})$ containing $\bar{M}$ and tangent to $h_u\left(N'^u_\epsilon\right)$ along $\bar{M}$ with trajectories in $W^u(\bar{M})$ approaching $\bar{M}$ as $t \rightarrow -\infty$. Moreover, the unstable manifold is persistent under perturbation in the sense that for any C^r vector field $f^{\mathrm{pert}}(x)$ C^1 θ-close to $f(x)$, with θ sufficiently small, there is a manifold $W^u\left(\bar{M}^{\mathrm{pert}}\right)$ overflowing invariant under $f^{\mathrm{pert}}(x)$ and C^r diffeomorphic to $W^u\left(\bar{M}\right)$.*

The proof of this theorem will proceed in stages much like the proof of the persistence theorem for overflowing invariant manifolds. We begin by constructing a *graph transform* and showing that it is well defined. We next show that this graph transform has a fixed point that corresponds to a Lipschitz unstable manifold of $\bar{M}$ for the time-T map generated by the flow. We then show that this manifold is invariant for all time. Next, we will turn to the question of differentiability and show that the Lipschitz manifold is actually C^r. Finally, we will show that the unstable manifold of $\bar{M}$ is persistent under perturbations.

The Graph Transform

In local coordinates, the image of a point on graph u is given by

$$
(x, u_i(x, z), z) \mapsto \left(f^0_{ij}(x, u_i(x, z), z), g^0_{ij}(x, u_i(x, z), z), h^0_{ij}(x, u_i(x, z), z) \right). \tag{4.29}
$$

Accordingly, the graph transform is defined by

$$
(Gu)_j (\xi_c, \xi_u) = g^0_{ij}(x, u_i(x, z), z), \tag{4.30}
$$

where

$$\xi_c = f_{ij}^0(x, u_i(x, z), z),$$
$$\xi_u = h_{ij}^0(x, u_i(x, z), z).$$

(Note: The subscript c denotes "center directions." A subscript T for "tangent directions" might have been better, however this could be confused with the T in the time-T map.) Before proceeding we must argue that this map is well defined. The argument is essentially the same as that given using Theorem 3.2.10 and Proposition 3.2.11. We summarize the necessary result in the following lemma.

Lemma 4.5.2 *Let*

$$\omega : N'_\epsilon \to N'^u_\epsilon,$$
$$(p, v^s, v^u) \to (p, v^u)$$

denote the fiber projection map and consider the map

$$\chi : N'^u_\epsilon \to N'^u_\epsilon,$$
$$(p, v^u) \to \omega \circ \phi_T \circ u \circ \omega \circ \phi_{-T}(p, v^u).$$

Then for δ and ϵ sufficiently small, and $u \in S_\delta$,

1. *χ is well defined for all $(p, v^u) \in \bar{U}_i^4 \times D^\epsilon$,*

2. *$\overline{\cup_{i=1}^s U_i^3 \times D^\epsilon} \subset \chi \left(\cup_{i=1}^s U_i^4 \times D^\epsilon \right) \subset \overline{\chi(\cup_{i=1}^s U_i^4 \times D^\epsilon)}$*
   $\subset \cup_{i=1}^s U_i^5 \times D^\epsilon.$

3. *Each point in $\cup_{i=1}^s U_i^3 \times D^\epsilon$ is the χ image of only one point in $\cup_{i=1}^s U_i^4 \times D^\epsilon$.*

Proof: The proof is identical to the proof of Proposition 3.2.11.   □

> **A Notational Point:** The notation $U_i^j \times D^\epsilon$ is not the most mathematically precise since N'^u_ϵ is not really covered by the sets $U_i^j \times D^\epsilon$ –
> this is just the notation for open sets V_i^j which cover N'^u_ϵ and (since is locally trivial) are diffeomorphic to $D^j \times D^\epsilon$.

Invariance of the graph implies that the graph transform has a fixed point which is expressed as

$$u_j \left(f_{ij}^0(x, u_i(x, z), z), h_{ij}^0(x, u_i(x, z), z) \right) = g_{ij}^0(x, u_i(x, z), z). \qquad (4.31)$$

We next want to show that the graph transform has a fixed point. However, we will not do this by proceeding directly with an analysis of the equations in this setup. Rather, we will show that the present setup can be recast into exactly the form of the equations studied when we were proving the existence of an overflowing invariant manifold for the perturbed vector field in Section 3.3.

We begin by defining

$$\bar{x} \equiv (x, z), \qquad \bar{g}_{ij}^0(\bar{x}, y) \equiv g_{ij}^0(x, y, z),$$

$$\bar{\xi} \equiv (\xi_c, \xi_u), \quad \bar{f}_{ij}^0(\bar{x}, y) = \left(f_{ij}^0(x, y, z), h_{ij}^0(x, y, z) \right) \tag{4.32}$$

with

$$\| \bar{x} \| \equiv \max \left(\| x \|, \| z \| \right),$$

$$\| \bar{\xi} \| \equiv \max \left(\| \xi_c \|, \| \xi_u \| \right), \tag{4.33}$$

$$\| \bar{f}_{ij}^0 \| \equiv \max \left(\| f_{ij}^0 \|, \| h_{ij}^0 \| \right).$$

In these new combined coordinates, $\bar{x}$, a section u is locally represented as

$$\bar{u}_i(\bar{x}) \equiv u_i(x, z). \tag{4.34}$$

We introduce the following shorthand notation:

$$\bar{A} \equiv D_1 \bar{g}_{ij}^0 \left(\bar{x}, \bar{u}_i(\bar{x}) \right) = \left(D_1 g_{ij}^0 \left(x, u_i(x, z), z \right), D_3 g_{ij}^0 \left(x, u_i(x, z), z \right) \right),$$

$$\bar{B} \equiv D_2 \bar{g}_{ij}^0 \left(\bar{x}, \bar{u}_i(\bar{x}) \right) = D_2 g_{ij}^0 \left(x, u_i(x, z), z \right),$$

$$\bar{C} \equiv D_1 \bar{f}_{ij}^0 \left(\bar{x}, \bar{u}_i(\bar{x}) \right) = D_{(1,3)} \left(f_{ij}^0 \left(x, u_i(x, z), z \right), h_{ij}^0 \left(x, u_i(x, z), z \right) \right),$$

$$\bar{E} \equiv D_2 \bar{f}_{ij}^0 \left(\bar{x}, \bar{u}_i(\bar{x}) \right) = \left(D_2 f_{ij}^0 \left(x, u_i(x, z), z \right), D_2 h_{ij}^0 \left(x, u_i(x, z), z \right) \right), \tag{4.35}$$

where $D_{(1,3)}(f_{ij}^0, h_{ij}^0)$ is defined as

$$\begin{pmatrix} D_1 f_{ij}^0 & D_3 f_{ij}^0 \\ D_1 h_{ij}^0 & D_3 h_{ij}^0 \end{pmatrix}.$$

Estimates for these quantities are given in the following lemma (which is analogous to Lemma 3.2.9).

Lemma 4.5.3 *For η, ϵ sufficiently small, and an appropriate constant $Q > 0$, we have the following estimates:*

1. $\| \bar{A} \|, \| \bar{B} \|, \| \bar{C} \|, \| \bar{E} \| < Q,$

2. $\| \bar{C}^{-1} \| < Q$,

3. $\| \bar{A} \| < \eta$, $\| \bar{B} \| < \frac{1}{2}$,

4. $\| \bar{B} \| \| \bar{C}^{-1} \| < \frac{1}{2}$.

Proof: Statement 1 follows immediately from the fact that $\bar{A}$, $\bar{B}$, $\bar{C}$, and $\bar{E}$ are C^{r-1} functions defined on a compact domain.

As for statement 2, we first note that $\| \bar{C}^{-1} \|$ exists. This follows from the overflowing invariance of $TM_2 \oplus N^u$ and the overflowing invariance (*under negative time*) of $TM_2 \oplus N^s$ under the linearized dynamics as well as from the fact that N'^s and N'^u can be selected arbitrarily C^1 close to N^s and N^u, respectively. But then $\bar{C}^{-1}$ is necessarily a C^{r-1} function and statement 2 follows.

Next we turn to the proof of statement 3. First, we can write

$$\| \bar{A} \| = \| D_1 \bar{g}_{ij}^0 \| = \| (D_1 g_{ij}^0, D_3 g_{ij}^0) \| \\ = \max \left(\| D_1 g_{ij}^0 \|, \| D_3 g_{ij}^0 \| \right) < \eta, \tag{4.36}$$

where we have used (4.23). Second, we have

$$\| \bar{B} \| = \| D_2 \bar{g}_{ij}^0 \| = \| D_2 g_{ij}^0 \| < \frac{1}{2}, \tag{4.37}$$

which follows from (4.16). This proves statement 3.

Finally, we turn to the proof of statement 4. We need to show that

$$\| \bar{B} \| \| \bar{C}^{-1} \| = \| D_2 \bar{g}_{ij}^0 \| \| \left(D_1 \bar{f}_{ij}^0 \right)^{-1} \| = \| D_2 g_{ij}^0 \| \| F^{-1} \| < \frac{1}{2} \tag{4.38}$$

holds with

$$F(x, y, z) = \begin{pmatrix} D_1 f_{ij}^0(x, y, z) & D_3 f_{ij}^0(x, y, z) \\ D_1 h_{ij}^0(x, y, z) & D_3 h_{ij}^0(x, y, z) \end{pmatrix}. \tag{4.39}$$

We define

$$a \equiv D_1 f_{ij}^0, \qquad b \equiv D_3 f_{ij}^0, \\ c \equiv D_1 h_{ij}^0, \qquad d \equiv D_3 h_{ij}^0, \tag{4.40}$$

where we have suppressed the arguments and indices for notational simplicity. Note that the GR-I estimates immediately give us

$$\| a \| < Q, \quad \| c \| < \eta, \quad \| a^{-1} \| < Q, \\ \| b \| < \eta, \quad \| d \| < Q, \quad \| d^{-1} \| < \frac{1}{4}. \tag{4.41}$$

(Note also that a and d are invertible matrices.) We will calculate F^{-1} by solving the following system of equations by elimination

$$\begin{aligned} am + bn &= p, \\ cm + dn &= q, \end{aligned} \tag{4.42}$$

where m, n, p, and q are of appropriate dimension. In this manner we easily obtain that

$$F^{-1} = \begin{pmatrix} \left(a - bd^{-1}c\right)^{-1} & -\left(a - bd^{-1}c\right)^{-1} bd^{-1} \\ \left(ca^{-1}b - d\right)^{-1} ca^{-1} & -\left(ca^{-1}b - d\right)^{-1} \end{pmatrix}. \tag{4.43}$$

We now estimate the norm of each individual entry of F^{-1}. We will use the estimates in (4.41) without explicit reference. Denoting the entries of the matrix F^{-1} by F_{ij}^{-1}, we can write

$$\| F_{11}^{-1} \| = \| \left(\mathbb{1} - a^{-1}bd^{-1}c\right)^{-1} a^{-1} \| \leq \| a^{-1} \| \left(1 + \mathcal{O}(\eta^2)\right)$$

$$= \| \left(D_1 f_{ij}^0\right)^{-1} \| \left(1 + \mathcal{O}(\eta^2)\right),$$

$$\| F_{12}^{-1} \| = \| -\left(\mathbb{1} - a^{-1}bd^{-1}c\right)^{-1} bd^{-1}a^{-1} \|$$

$$\leq \| a^{-1} \| \| b \| \| d^{-1} \| \left(1 + \mathcal{O}(\eta^2)\right)$$

$$< \tfrac{1}{4} Q\eta \left(1 + \mathcal{O}(\eta^2)\right) = \mathcal{O}(\eta),$$

$$\| F_{21}^{-1} \| = \| \left(d^{-1}ca^{-1}b - \mathbb{1}\right)^{-1} d^{-1}ca^{-1} \| \tag{4.44}$$

$$\leq \| a^{-1} \| \| c \| \| d^{-1} \| \left(1 + \mathcal{O}(\eta^2)\right)$$

$$< \tfrac{1}{4} Q\eta \left(1 + \mathcal{O}(\eta^2)\right) = \mathcal{O}(\eta),$$

$$\| F_{22}^{-1} \| = \| -\left(d^{-1}ca^{-1}b - \mathbb{1}\right)^{-1} d^{-1} \| \leq \| d^{-1} \| \left(1 + \mathcal{O}(\eta^2)\right)$$

$$< \tfrac{1}{4} \left(1 + \mathcal{O}(\eta^2)\right).$$

To be consistent with the maximum norms used for the coordinate $\bar{x}$, we calculate the norm of F^{-1} as

$$\| F^{-1} \| = \max_{1 \leq i,j \leq 2} \left(\| F_{ij}^{-1} \|\right) < 2 \max \left(\| \left(D_1 f_{ij}^0\right)^{-1} \|, \tfrac{1}{4}\right) \tag{4.45}$$

which holds based on the above estimates for η sufficiently small. Then (4.38), (4.45), and the GR-I estimates give

$$\| \bar{B} \| \, \| \bar{C}^{-1} \| = \| D_2 g_{ij}^0 \| \, \| F^{-1} \|$$

$$< 2 \max \left(\| D_2 g_{ij}^0 \| \, \| \left(D_1 f_{ij}^0 \right)^{-1} \|, \tfrac{1}{4} \, \| D_2 g_{ij}^0 \| \right) \qquad (4.46)$$

$$< 2 \max \left(\tfrac{1}{4}, \tfrac{1}{4} \cdot \tfrac{1}{2} \right) = \tfrac{1}{2},$$

which completes the proof of the lemma.   □

Next we turn to the question of existence of the unstable manifold of the overflowing invariant manifold $\bar{M}$. However, first note that in the "combined coordinates" the image of a point on graph u under the time-T map is given as

$$\left(\bar{x}, \bar{u}_i(\bar{x}) \right) \mapsto \left(\bar{f}_{ij}^0 \left(\bar{x}, \bar{u}_i(\bar{x}) \right), \bar{g}_{ij}^0 \left(\bar{x}, \bar{u}_i(\bar{x}) \right) \right). \qquad (4.47)$$

The graph transform G is now written as

$$\left(\overline{Gu} \right)_j (\bar{\xi}) = \bar{g}_{ij}^0 \left(\bar{x}, \bar{u}_i(\bar{x}) \right), \qquad (4.48)$$

where

$$\bar{\xi} = \bar{f}_{ij}^0 \left(\bar{x}, \bar{u}_i(\bar{x}) \right).$$

The graph of u is a fixed point of the graph transform if

$$\bar{u}_j \left(\bar{f}_{ij}^0 \left(\bar{x}, \bar{u}_i(\bar{x}) \right) \right) = \bar{g}_{ij}^0 \left(\bar{x}, \bar{u}_i(\bar{x}) \right). \qquad (4.49)$$

One can easily see that the expressions in (4.35) and (4.47)–(4.49) are exactly of the form of the expressions (3.62) in Section 3.3 and (3.55)–(3.57) in Section 3.2.7. Moreover, $\bar{A}$, $\bar{B}$, $\bar{C}$, and $\bar{E}$ satisfy the same estimates as A, B, C, and E (cf. the lemma above as well as Lemma 3.2.9). Thus, although the graph transform is now defined on a space with a different structure, its coordinate expressions are of the same type as previously studied and their derivatives obey the same estimates. One can use this analogy to establish the following sequence of results.

Proposition 4.5.4 *For ϵ, δ, and η sufficiently small,*

$$G : S_\delta \to S_\delta.$$

Proof: By analogy, using Proposition 3.3.2, we have

$$\| \left(\overline{Gu} \right)_j (\bar{\xi}) - \left(\overline{Gu} \right)_j (\bar{\xi}') \| \leq \delta \, \| \bar{\xi} - \bar{\xi}' \| .$$

Since

$$\| \bar{\xi} - \bar{\xi}' \| = \max \left(\| \xi_c - \xi_c' \|, \| \xi_u - \xi_u' \| \right),$$

we have

$$
\begin{aligned}
\| (Gu)_j (\xi_c, \xi_u) - (Gu)_j (\xi_c', \xi_u') \| &\leq \delta \max \left(\| \xi_c - \xi_c' \|, \| \xi_u - \xi_u' \| \right) \\
&\leq \delta \left(\| \xi_c - \xi_c' \| + \| \xi_u - \xi_u' \| \right).
\end{aligned}
$$

Hence, the proposition is proved. □

Proposition 4.5.5 *For ϵ, δ, and η sufficiently small, G is a contraction on S_δ in the C^0 norm.*

Proof: Let $(\xi_c, \xi_u) \in \left(\sigma_j \times \tau_j^u \right) \left(N_\epsilon' |_{U_j^3} \right)$ be given. We choose $u, u' \in S_\delta$, i, x, x' and z, z' such that

$$
\begin{aligned}
\xi_c &= f_{ij}^0 \left(x, u_i(x, z), z \right) = f_{ij}^0 \left(x', u_i(x', z'), z' \right), \\
\xi_u &= h_{ij}^0 \left(x, u_i(x, z), z \right) = h_{ij}^0 \left(x', u_i(x', z'), z' \right)
\end{aligned}
$$

or, equivalently,

$$\xi = \bar{f}_{ij}^0 \left(\bar{x}, \bar{u}_i(\bar{x}) \right) = \bar{f}_{ij}^0 \left(\bar{x}', \bar{u}_i(\bar{x}') \right).$$

By analogy, using Proposition 3.3.3, we have

$$\| \overline{(Gu)}_j (\bar{\xi}) - \overline{(Gu')}_j (\bar{\xi}') \| \leq \delta \| \bar{u}_i - \bar{u}_i' \|_0$$

or, equivalently,

$$\| (Gu)_j (\xi_c, \xi_u) - (Gu')_j (\xi_c, \xi_u) \| \leq \delta \| u_i - u_i' \|_0 .$$

Hence, the proposition is proved. □

Corollary 4.5.6 *There is a unique $u \in S_\delta$ such that $\phi_t (\text{graph } u) \supset \text{graph } u$ for all $t > 0$.*

Proof: The proof is identical to the proof of Corollary 3.3.4. □

This completes the proof of the existence of $W^u \left(\bar{M} \right)$.

4.5.1 DIFFERENTIABILITY OF $W^u(\bar{M})$

The proof that $W^u(\bar{M})$ is C^r goes exactly as the proof of the differentiability of the perturbed overflowing invariant manifold given earlier. The derivative of u_i, Du_i, is denoted by

$$Du_i = (D_1 u_i, D_3 u_i) \equiv (v_i, w_i) \in C^0(\mathcal{D}^3 \times \mathcal{D}^\epsilon, \mathrm{L}(\mathrm{IR}^{n-(s+u)} \times \mathrm{IR}^u, \mathrm{IR}^s)),$$

and the derivative of u, Du, is denoted by the s maps

$$Du \equiv (v, w) \equiv ((v_1, \ldots, v_s), (w_1, \ldots, w_s)),$$

where

$$Du \in [C^0(\mathcal{D}^3 \times \mathcal{D}^\epsilon, \mathrm{L}(\mathrm{IR}^{n-(s+u)} \times \mathrm{IR}^u, \mathrm{IR}^s))]^s.$$

We define the norm of Du as

$$\| Du \| \equiv \max_i \sup_{(x,z) \in \mathcal{D}^3 \times \mathcal{D}^\epsilon} \| Du_i \|.$$

Notation: The lowercase s is used to denote both the dimension of the unstable manifold and the number of charts. Certainly, they need not be the same and the difference in usage should be clear from the context.

First, we derive a functional equation that the first derivative of $W^u(\bar{M})$ must satisfy, if it exists. From (4.31), in local coordinate charts, u must satisfy

$$u_j(\xi_c, \xi_u) = g_{ij}^0(x, u_i(x, z), z), \tag{4.50}$$

where

$$\xi_c = f_{ij}^0(x, u_i(x, z), z), \tag{4.51}$$
$$\xi_u = h_{ij}^0(x, u_i(x, z), z). \tag{4.52}$$

In the combined coordinates, (4.50)–(4.52) simplifies to

$$\bar{u}_j(\bar{\xi}) = \bar{g}_{ij}^0(\bar{x}, \bar{u}_i(\bar{x}))$$

with

$$\bar{\xi} = \bar{f}_{ij}^0(\bar{x}, \bar{u}_i(\bar{x})).$$

Formal differentiation of $\bar{u}_j(\bar{\xi})$ with respect to x gives

$$D\bar{u}_j(\bar{\xi}) \equiv \bar{v}_j(\bar{\xi}) = [\bar{A} + \bar{B}\bar{v}_i(\bar{x})][\bar{C} + \bar{E}\bar{v}_i(\bar{x})]^{-1}, \tag{4.53}$$

in complete analogy with Eq. (3.92). From this point on, every step in the proof of differentiability is the same as in Section 3.3.1 since the estimates for $\bar{A}$, $\bar{B}$, $\bar{C}$, and $\bar{E}$ are the same as for A, B, C, and E.

TANGENCY

In order to show that $W^u(\bar{M})$ is tangent to $h_u\left(N'^u_\epsilon\right)$ along $\bar{M}$ it suffices to show the following:

1. $u_j(x,0) = 0$, for $j = 1, \ldots, s$,

2. $D_3 u_j(x,0) = 0$, for $j = 1, \ldots, s$.

To prove the first part we argue by contradiction. Suppose $u_j(x,0) \neq 0$ for some j. Since ϵ is arbitrary, it can be chosen small enough such that graph u is not contained in $h\left(N'_\epsilon\right)$. However, this contradicts the fact that u is the unique fixed point of the graph transform.

For the second part, we note that by construction we have

$$\| Du \| < \delta. \tag{4.54}$$

Choosing ϵ and δ small enough, the graph transform always has a fixed point obeying (4.54). Thus, $\| Du \| \mid_{\bar{M}}$ must be smaller than any positive number; hence, $\| Du \| \mid_{\bar{M}} = 0$. But this implies that $\| D_3 u_j(x,0) \| = 0$, from which we conclude that $D_3 u_j(x,0) = 0$.

4.5.2 $W^u\left(\bar{M}\right)$ SATISFIES THE HYPOTHESES OF THE PERSISTENCE THEOREM FOR OVERFLOWING INVARIANT MANIFOLDS

We have constructed the overflowing invariant manifold $W^u(\bar{M})$. Strictly speaking, we have shown the existence of a local unstable manifold $W^u_{\text{loc}}(\bar{M})$ which is C^0 ϵ-close to the manifold $h_u(N'^u_\epsilon|_{\cup_i U^6_i})$. We want to argue that the manifold that we have constructed satisfies the hypotheses of the persistence theorem for overflowing invariant manifolds (Theorem 3.3.1). In this regard, the superscript 0 will refer to quantities related to the unperturbed vector field and the superscript pert will refer to quantities related to the perturbed vector field.

Since u is Lipschitz with global Lipschitz constant δ, we can write

$$\| Du \| \leq \delta.$$

Since $W^u_{\text{loc}}(\bar{M})$ is the graph of u over $N'^u_\epsilon|_{\cup_i U^6_i}$, this last inequality implies that $W^u_{\text{loc}}(\bar{M})$ and $h_u(N'^u_\epsilon|_{\cup_i U^6_i})$ are, in fact, C^1 α-close to $\alpha = \max(\epsilon, \delta)$. This means that in a coordinate system fixed to $W^u_{\text{loc}}(\bar{M})$, the coordinate expressions for the time-T map (i.e., $\bar{f}^0_{ij}$, $\bar{g}^0_{ij}$), and the relevant expressions involving the derivatives of the time-T map (i.e., $\bar{A}^0$, $\bar{B}^0$, $\bar{C}^0$, and $\bar{E}^0$) are uniformly C^0 α-close to the same quantities computed in the same coordinate system attached to $h_u(N'^u_\epsilon|_{\cup_i U^6_i})$. This implies that for ϵ and δ

sufficiently small, we have

$$\| \bar{A}^0 \|, \| \bar{B}^0 \|, \| \bar{C}^0 \|, \| \bar{E}^0 \| < Q,$$

$$\| (\bar{C}^0)^{-1} \| < Q,$$

$$\| \bar{A}^0 \| < \eta, \quad \| \bar{B}^0 \| < \tfrac{1}{2},$$

$$\| \bar{B}^0 \| \, \| (\bar{C}^0)^{-1} \| < \tfrac{1}{2},$$

which one obtains from (4.15). If we now compute the same quantities for a C^1 θ-close flow $\phi_t^{\mathrm{pert}}(\cdot)$, then for θ small enough one still has

$$\| \bar{A}^{\mathrm{pert}} \|, \| \bar{B}^{\mathrm{pert}} \|, \| \bar{C}^{\mathrm{pert}} \|, \| \bar{E}^{\mathrm{pert}} \| < Q,$$

$$\| (\bar{C}^{\mathrm{pert}})^{-1} \| < Q,$$

$$\| \bar{A}^{\mathrm{pert}} \| < \eta, \quad \| \bar{B}^{\mathrm{pert}} \| < \tfrac{1}{2},$$

$$\| \bar{B}^{\mathrm{pert}} \| \| (\bar{C}^{\mathrm{pert}})^{-1} \| < \tfrac{1}{2}.$$

But in that case, one can repeat the construction in Section 3.3 to prove the existence of a local overflowing invariant manifold $W_{\mathrm{loc}}^{u,\mathrm{pert}}$ using the analogy we utilized to prove the existence of $W_{\mathrm{loc}}^u(\bar{M}) \equiv W_{\mathrm{loc}}^{u,0}$. The advantage now is that one does not have to compute generalized Lyapunov-type numbers for $W_{\mathrm{loc}}^{u,0}$ to obtain the above estimates which are necessary to solve the persistence problem.

5

Foliations of Unstable Manifolds

Now we consider the situation where $W^u(\bar{M})$ is *foliated* or *fibered* by submanifolds corresponding to orbits that approach each other (in negative time) at the fastest rate. These will also turn out to be the initial conditions for trajectories in $W^u(\bar{M})$ that approach the same trajectory in $\bar{M}$ (in negative time). The geometrical setup for this situation, in the sense of the construction of local coordinates near $\bar{M}$ and $W^u(\bar{M})$, is the same as that for the proof of the unstable manifold theorem for overflowing invariant manifolds.

5.1   Generalized Lyapunov-Type Numbers

Suppose we have the following continuous splitting

$$T\mathbb{R}^n|_{M_2} = TM_2 \oplus N^s \oplus N^u$$

and the associated projections

$$\Pi^u \quad : \quad T\mathbb{R}^n|_{M_2} \to N^u, \tag{5.1}$$
$$\Pi^s \quad : \quad T\mathbb{R}^n|_{M_2} \to N^s. \tag{5.2}$$

For notational purposes we define $N \equiv N^s \oplus N^u$. We assume that the subbundles $TM_2 \oplus N^u$ and $TM_2 \oplus N^s$ are each invariant under $D\phi_t$ for all $t < 0$ (i.e., overflowing invariant). Growth rates of vectors in these subbundles under the linearized dynamics are characterized by generalized Lyapunov-type numbers as before. For a point $p \in M_2$ we consider the following *nonzero* vectors:

$$u_0 \quad \in \quad N_p^u,$$
$$w_0 \quad \in \quad N_p^s,$$
$$v_0 \quad \in \quad T_p M_2$$

and

$$u_{-t} \quad = \quad \Pi^u D\phi_{-t}(p)u_0,$$

$$w_{-t} = \Pi^s D\phi_{-t}(p)w_0,$$
$$v_{-t} = D\phi_{-t}(p)v_0.$$

We define the following generalized Lyapunov-type numbers:

$$\lambda^u(p) = \inf\left\{a : \left(\frac{\|u_{-t}\|}{\|u_0\|}\right)/a^t \to 0 \quad \text{as} \quad t\uparrow\infty \quad \forall u_0 \in N_p^u\right\}. \quad (5.3)$$

$$\sigma^{cu}(p) = \inf\left\{\rho : ((\|u_{-t}\|/\|v_{-t}\|)/(\|u_0\|/\|v_0\|))/\rho^t \to 0\right.$$
$$\left.\text{as} \quad t\uparrow\infty \quad \forall v_0 \in T_pM_2, \quad u_0 \in N_p^u\right\}. \quad (5.4)$$

$$\sigma^{su}(p) = \inf\left\{\rho : ((\|u_{-t}\|/\|w_{-t}\|)/(\|u_0\|/\|w_0\|))/\rho^t \to 0\right.$$
$$\left.\text{as} \quad t\uparrow\infty \quad \forall w_0 \in N_p^s, \quad u_0 \in N_p^u\right\}. \quad (5.5)$$

Following the same calculations as in Lemma 3.1.1, we can show that these expressions are equal to the following:

$$\lambda^u(p) = \overline{\lim_{t\to\infty}} \parallel \Pi^u D\phi_{-t}(p)|_{N_p^u} \parallel^{\frac{1}{t}}, \quad (5.6)$$

$$\sigma^{cu}(p) = \overline{\lim_{t\to\infty}} \parallel D\phi_t|_{M_2}(\phi_{-t}(p)) \parallel^{\frac{1}{t}} \parallel \Pi^u D\phi_{-t}(p)|_{N_p^u} \parallel^{\frac{1}{t}}, \quad (5.7)$$

$$\sigma^{su}(p) = \overline{\lim_{t\to\infty}} \parallel \Pi^u D\phi_{-t}(p)|_{N_p^u} \parallel^{\frac{1}{t}} \parallel \Pi^s D\phi_t(\phi_{-t}(p))|_{N_p^s} \parallel^{\frac{1}{t}}. \quad (5.8)$$

Lemma 3.1.2 (constant on orbits) and Proposition 3.1.3 (independent of metric) also apply to these generalized Lyapunov-type numbers. The proofs involve identical calculations.

The geometrical interpretation of these generalized Lyapunov-type numbers should be clear from the earlier discussion. Moreover, we have the following *uniformity lemma.*

Lemma 5.1.1 (Uniformity Lemma for Foliations)

1. *Suppose*

$$\parallel \Pi^u D\phi_{-t}(p)|_{N_p^u} \parallel /a^t \to 0 \quad \text{as} \quad t\uparrow\infty, \quad \forall p \in \bar{M}_1.$$

   Then there are constants $\hat{a} < a$ and C such that

$$\parallel \Pi^u D\phi_{-t}(p)|_{N_p^u} \parallel < C\hat{a}^t \quad \forall p \in \bar{M}_1 \quad \text{and} \quad t \geq 0.$$

2. Suppose that

$$\| \ \Pi^u D\phi_{-t}(p)|_{N_p^u} \ \| \ \| \ D\phi_t|_{M_2} \left(\phi_{-t}(p)\right) \ \| \ /\rho^t \to 0$$

$$as \quad t \uparrow \infty, \quad \forall p \in \bar{M}_1.$$

Then there are constants $\hat{\rho} < \rho$ and C such that

$$\| \ \Pi^u D\phi_{-t}(p)|_{N_p^u} \ \| \ \| \ D\phi_t|_{M_2} \left(\phi_{-t}(p)\right) \ \| < C\hat{\rho}^t,$$

$$\forall p \in \bar{M}_1 \quad and \quad t \geq 0.$$

3. Suppose that

$$\| \ \Pi^s D\phi_t \left(\phi_{-t}(p)\right)|_{N_p^s} \ \| \ \| \ \Pi^u D\phi_{-t}(p)|_{N_p^u} \ \| \ /\rho^t \to 0$$

$$as \quad t \uparrow \infty, \forall p \in \bar{M}_1.$$

Then there are constants $\hat{\rho} < \rho$ and C such that

$$\| \ \Pi^s D\phi_t \left(\phi_{-t}(p)\right)|_{N_p^s} \ \| \ \| \ \Pi^u D\phi_{-t}(p)|_{N_p^u} \ \| < C\hat{\rho}^t,$$

$$\forall p \in \bar{M}_1 \quad and \quad t \geq 0.$$

4. If $\lambda^u(p) < a \leq 1$, $\sigma^{cu}(p) < \rho \leq 1$, and $\sigma^{su}(p) < \rho \leq 1$, $\forall p \in \bar{M}_1$, then

$$\| \ \Pi^u D\phi_{-t}(p)|_{N_p^u} \ \| \to 0 \quad as \quad t \uparrow \infty,$$

$$\| \ D\phi_t|_{M_1} \left(\phi_{-t}(p)\right) \ \| \ \| \ \Pi^u D\phi_{-t}(p)|_{N_p^u} \ \| \to 0 \quad as \quad t \uparrow \infty,$$

and

$$\| \ \Pi^s D\phi_t \left(\phi_{-t}(p)\right)|_{N_p^s} \ \| \ \| \ \Pi^u D\phi_{-t}(p)|_{N_p^u} \ \| \to 0 \quad as \quad t \uparrow \infty$$

uniformly $\forall p \in \bar{M}_1$.

5. $\lambda^u(\cdot)$, $\sigma^{cu}(\cdot)$, and $\sigma^{su}(\cdot)$ attain their suprema on M.

Proof: The proof proceeds exactly as in the proof of the uniformity lemma given earlier. □

5.2   Local Coordinates Near M

The construction here, and subsequent notation, is exactly the same as for the proof of the unstable manifold theorem.

5.3   The GR-I Estimates

Henceforth we will assume that

$$\boxed{\lambda^u(p) < 1,\ \sigma^{cu}(p) < 1,\ \text{and}\ \sigma^{su}(p) < 1\ \text{for every}\ p \in \bar{M}_1.}$$

It follows from this assumption and the uniformity lemma that there is a $T > 1$ such that

$$\| \Pi'^u D\phi_{-T}(p)|_{N_p^u} \| < \min\left\{\tfrac{a^T}{8}, \tfrac{a^T}{8\mathcal{K}}\right\} < \min\left\{\tfrac{1}{8}, \tfrac{1}{8\mathcal{K}}\right\},$$

$$c\, \| D\phi_T|_{M_2}\, (\phi_{-T}(p)) \| \, \| \Pi'^u D\phi_{-T}(p)|_{N_p^u} \| \quad < \quad \min\left\{\tfrac{\rho^T}{8}, \tfrac{\rho^T}{8\mathcal{K}}\right\}$$

$$< \quad \min\left\{\tfrac{1}{8}, \tfrac{1}{8\mathcal{K}}\right\}, \qquad (5.9)$$

$$\| \Pi'^u D\phi_{-T}(p)|_{N_p^u} \| \, \| \Pi'^s D\phi_T\, (\phi_{-T}(p))|_{N_p^s} \| < \quad \min\left\{\tfrac{\rho^T}{8}, \tfrac{\rho^T}{8\mathcal{K}}\right\}$$

$$< \quad \min\left\{\tfrac{1}{8}, \tfrac{1}{8\mathcal{K}}\right\},$$

where

$$\mathcal{K} \equiv \| D_2 f_{ij}^0 \|,$$

and we are using the perturbed normal bundle which gains a derivative as the generalized Lyapunov-type numbers are independent of the projection. We fix this T for the rest of our arguments. Exactly as in the case for the unstable manifold theorem, we translate these estimates into the following local coordinate expressions for the time-T map. We skip some of the preliminary details and give the results:

$$\| \left(D_3 h_{ij}^0\right)^{-1} \| < \min\left\{\tfrac{a^T}{8}, \tfrac{a^T}{8\mathcal{K}}\right\} < \min\left\{\tfrac{1}{8}, \tfrac{1}{8\mathcal{K}}\right\},$$

$$\| D_1 f_{ij}^0 \| \, \| \left(D_3 h_{ij}^0\right)^{-1} \| < \min\left\{\tfrac{\rho^T}{8}, \tfrac{\rho^T}{8\mathcal{K}}\right\} < \min\left\{\tfrac{1}{8}, \tfrac{1}{8\mathcal{K}}\right\}, \qquad (5.10)$$

$$\| D_2 g_{ij}^0 \| \, \| \left(D_3 h_{ij}^0\right)^{-1} \| < \min\left\{\tfrac{\rho^T}{8}, \tfrac{\rho^T}{8\mathcal{K}}\right\} < \min\left\{\tfrac{1}{8}, \tfrac{1}{8\mathcal{K}}\right\}.$$

The following estimates also follow from the setup for the unstable manifold theorem (the GR-I estimates), we restate them here for ease of reference:

$$\| g_{ij}^0(x, y, z) \| < \eta,$$

$$\| h_{ij}^0(x, y, z) \| < \eta,$$

$$\| D_1 g_{ij}^0(x, y, z) \| < \eta,$$

$$\| D_1 h_{ij}^0(x, y, z) \| < \eta, \tag{5.11}$$

$$\| D_2 h_{ij}^0(x, y, z) \| < \eta,$$

$$\| D_3 g_{ij}^0(x, y, z) \| < \eta,$$

$$\| D_3 f_{ij}^0(x, y, z) \| < \eta.$$

Recall, that these estimates hold *for a given $\eta > 0$ and ϵ sufficiently small.*

5.4   The Space of Families of u-Dimensional Lipschitz Surfaces

Consider families of maps of the form

$$f_1(\cdot; p) : \mathcal{D}^\epsilon \to \mathbb{R}^{n-(s+u)}, \tag{5.12}$$

where, for the moment, p is to be regarded as a parameter satisfying $\sigma_i(p) = f_1(0; p)$, for $p \in U_i$, $i = 1, \ldots, s$. We define the Lipshitz constant of $f_1(\cdot; p)$ as follows:

$$\text{Lip } f_1 \equiv \sup \left\{ \frac{\| f_1(z; p) - f_1(z'; p) \|}{\| z - z' \|}, \, z, z' \in \mathcal{D}^\epsilon, z \neq z' \right\}. \tag{5.13}$$

Let $S_{\delta,1}^p$ denote the set of functions of the form (5.12) with Lipschitz constants less than or equal to δ. We define a metric on $S_{\delta,1}^p$ as follows:

$$d_1^p(f_1, \hat{f}_1) = \sup \left\{ \frac{\| f_1(z; p) - \hat{f}_1(z; p) \|}{\| z \|}, z \in \mathcal{D}^\epsilon \right\}, \tag{5.14}$$

where $f_1(\cdot; p), \hat{f}_1(\cdot; p) \in S_{\delta,1}^p$. Note that $f_1(0; p) = \hat{f}_1(0; p)$. This metric is not the metric derived from the C^0 norm, as was used in the proof of the persisting overflowing invariant manifold theorem and the unstable manifold theorem.

Similarly, consider families of maps of the form

$$f_2(\cdot;p) : \mathcal{D}^\epsilon \to \mathbb{R}^s, \qquad f_2(0;p) = 0, \tag{5.15}$$

where p is to be regarded as a parameter satisfying $\sigma_i(p) = f_1(0;p)$, for $p \in U_i$, $i = 1,\ldots,s$. We define the Lipshitz constant of $f_2(\cdot;p)$ in the same manner as follows:

$$\text{Lip}\, f_2 \equiv \sup\left\{ \frac{\|\, f_2(z;p) - f_2(z';p)\,\|}{\|\, z - z'\,\|}, \; z, z' \in \mathcal{D}^\epsilon, z \neq z' \right\}. \tag{5.16}$$

Let $S^p_{\delta,2}$ denote the set of functions with Lipschitz constants less than or equal to δ. We define a metric on $S^p_{\delta,2}$ as

$$d^p_2(f_2, \hat{f}_2) = \sup\left\{ \frac{\|\, f_2(z;p) - \hat{f}_2(z;p)\,\|}{\|\, z\,\|}, \; z \in \mathcal{D}^\epsilon \right\}, \tag{5.17}$$

where $f_2(\cdot;p), \hat{f}_2(\cdot;p) \in S^p_{\delta,2}$.

Now let S^p_δ denote the families of functions of the form

$$\mathcal{F}_p : \mathcal{D}^\epsilon \; \to \; \mathbb{R}^{n-(s+u)} \times \mathbb{R}^s, \tag{5.18}$$

$$z \; \mapsto \; (f_1(z;p), f_2(z;p)) \equiv \mathcal{F}_p(z). \tag{5.19}$$

We define the Lipschitz constant of $\mathcal{F}_p(z)$ as follows:

$$\text{Lip}\, \mathcal{F}_p \equiv \max\left\{ \text{Lip}\, f_1, \text{Lip}\, f_2 \right\}. \tag{5.20}$$

We define a metric on S^p_δ as

$$d^p\left(\mathcal{F}_p, \hat{\mathcal{F}}_p\right) \equiv \max\left\{ d^p_1\left(f_1, \hat{f}_1\right), d^p_2\left(f_2, \hat{f}_2\right) \right\}. \tag{5.21}$$

With this metric, S^p_δ is a complete metric space.

The graphs of the functions $\mathcal{F}_p$ are u dimensional and, when viewed as surfaces in our local coordinates, intersect M in the point $\sigma_i(p) = f_1(0;p)$, for $p \in U_i$, $i = 1,\ldots,s$. We will use these spaces in contraction mapping arguments as earlier in order to show the existence of a foliation of the unstable manifold. We illustrate the geometry in Fig. 5.1.

5.5 Sequences of Contraction Maps

The contraction map setup for the foliation theorem will be different from that of the persistence theorem for overflowing invariant manifolds and the unstable manifold theorem. Therefore, we will first give a description in a general setting that is inspired by the work of Silnikov [1967]. Let

$$\{X_i, d_i(\cdot,\cdot)\}, \qquad i = 0, 1, 2, \ldots,$$

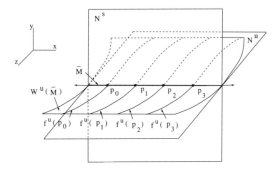

FIGURE 5.1. Geometry Associated with the Families of Maps—a Mnemonic Illustration.

denote an infinite sequence of complete metric spaces, with maps A_i acting on each X_i, where the following conditions hold for all i:

$$A_i X_i \subset X_{i-1}, \quad i = 1, \ldots,$$

$$d_{i-1}(A_i x_i, A_i x_i') \le q d_i(x_i, x_i'), \qquad x_i, x_i' \in X_i, \qquad (5.22)$$

$$\sup_{x_i, x_i' \in X_i} d_i(x_i, x_i') < C < \infty,$$

where A_i is a map of X_i into X_{i-1}, and C and q are positive constants. Note that by (5.22) each A_i is Lipschitz with Lipschitz constant q. Consider the product space

$$X \equiv \prod_{i=0}^{+\infty} X_i,$$

where elements of X are infinite sequences denoted by

$$x \equiv (x_0, x_1, x_2, \ldots).$$

We define a metric on X as

$$d(x, x') = \sup_{0 \le i < \infty} d_i(x_i, x_i').$$

X is a complete metric space with this metric. We define a map on X as

$$\mathcal{A} : X \quad \rightarrow \quad X, \qquad (5.23)$$

$$x \quad \mapsto \quad \mathcal{A}x \equiv \bar{x}, \qquad (5.24)$$

where

$$\bar{x}_i \equiv A_{i+1} x_{i+1}, \qquad i = 0, 1, \dots.$$

We have the following proposition.

Proposition 5.5.1 $\mathcal{A}$ *is a contraction map if $q < 1$.*

Proof: See Silnikov [1967].   □

By the contraction mapping theorem, $\mathcal{A}$ has a unique fixed point. We want to consider the nature of this fixed point. Suppose $\mathcal{A}x = x$, then we have

$$x_i = A_{i+1} x_{i+1}.$$

So we see that a fixed point of $\mathcal{A}$ is an infinite sequence having the property that the ith element is the image of the $(i+1)^{\text{th}}$ element under the map A_{i+1}.

5.6   The Theorem on Foliations of Unstable Manifolds

Theorem 5.6.1 *Suppose $\dot{x} = f(x)$ is a C^r vector field on $\mathbb{R}^n$, $r \geq 1$. Let $\bar{M} \equiv M \cup \partial M$ be a C^r compact connected manifold with boundary, overflowing invariant under the vector field $f(x)$. Suppose $\lambda^u(p) < 1$, $\sigma^{cu}(p) < 1$, and $\sigma^{su}(p) < 1$ for every $p \in \bar{M}_1$. Then there exists a $n - (s + u)$-parameter family $\mathcal{F}^u = \cup_{p \in M} f^u(p)$ of u-dimensional surfaces $f^u(p)$ (with boundary), such that the following hold:*

1. *$\mathcal{F}^u$ is a negatively invariant family, i.e., $\phi_{-t}\left(f^u(p)\right) \subset f^u\left(\phi_{-t}(p)\right)$ for any $t \geq 0$ and $p \in M$.*

2. *The u-dimensional surfaces $f^u(p)$ are C^r.*

3. *$f^u(p)$ is tangent to $h_u\left(N_p'^u\right)$ at p.*

4. *There exist $C_u, \lambda_u > 0$ such that if $q \in f^u(p)$, then*

$$\| \phi_{-t}(q) - \phi_{-t}(p) \| < C_u e^{-\lambda_u t}$$

   for any $t \geq 0$.

5. *Suppose $q \in f^u(p)$ and $q' \in f^u(p')$. Then*

$$\frac{\| \phi_{-t}(q) - \phi_{-t}(p) \|}{\| \phi_{-t}(q') - \phi_{-t}(p) \|} \to 0 \qquad as \quad t \uparrow \infty,$$

   unless $p = p'$.

6. $f^u(p) \cap f^u(p') = \emptyset$, unless $p = p'$.

7. If the hypotheses of the unstable manifold theorem hold, i.e., if additionally $\nu^s(p) < 1$ and $\sigma^s(p) < \frac{1}{r}$ for every $p \in \bar{M}_1$, then the u-dimensional surfaces $f^u(p)$ are C^r with respect to the basepoint p.

8. $\mathcal{F}^u = W^u_{loc}(M)$.

The most extensive part the proof involves part 1. For this we must construct a graph transform, show that the graph transform has a fixed point which is the negatively invariant family of surfaces, and show that the surfaces have the desired differentiability properties. This is established through a series of lemmas and propositions as in the proof of the persistence of overflowing invariant manifolds and in the proof of the unstable manifold theorem for overflowing invariant manifolds.

5.6.1   THE PROOF OF THE THEOREM

THE GRAPH TRANSFORM

Consider the map

$$
\begin{pmatrix} f_1(z;p) \\ f_2(z;p) \\ z \end{pmatrix} \mapsto
\begin{pmatrix} f^0_{ij}\left(f_1(z;p), f_2(z;p), z\right) \\ g^0_{ij}\left(f_1(z;p), f_2(z;p), z\right) \\ h^0_{ij}\left(f_1(z;p), f_2(z;p), z\right) \end{pmatrix}. \tag{5.25}
$$

Neglecting, for the moment, the issue of whether or not this map is well defined, if graph $\mathcal{F}_p(z)$ is contained in an invariant family, then we have

$$
f_1\left(h^0_{ij}\left(f_1(z;p), f_2(z;p), z\right); p'\right) = f^0_{ij}\left(f_1(z;p), f_2(z;p), z\right),
$$
$$
f_2\left(h^0_{ij}\left(f_1(z;p), f_2(z;p), z\right); p'\right) = g^0_{ij}\left(f_1(z;p), f_2(z;p), z\right),
$$

(5.26)

where

$$
\sigma_j(p') = f^0_{ij}(x,0,0), \quad f_1(0;p) = x, \quad f_2(0;p) = 0.
$$

Hence, we can define the following graph transform:

$$
[G_p(\mathcal{F}_p)](\xi, p') = \left(f^0_{ij}\left(f_1(z;p), f_2(z;p), z\right), g^0_{ij}\left(f_1(z;p), f_2(z;p), z\right)\right),
$$

where

$$
\xi = h^0_{ij}\left(f_1(z;p), f_2(z;p), z\right), \quad \sigma_j(p') = f^0_{ij}(x,0,0).
$$

Now we show how the general setup for the sequences of contraction maps can be related to the graph transform. Let $p \in M$ and let

$$\{p \equiv p_0, p_1, p_2, \ldots\}$$

denote a sequence defined as

$$p_i = \phi_{-T}(p_{i-1}), \quad i \geq 1.$$

Then we can consider the infinite sequence of complete metric spaces

$$\{S_\delta^{p_i}, d^{p_i}\}.$$

We form the product space

$$S_\delta \equiv \prod_{i=0}^{\infty} S_\delta^{p_i},$$

with the metric on S_δ given by

$$d(\mathcal{F}, \hat{\mathcal{F}}) = \sup_{0 \leq i < \infty} d^{p_i}\left(\mathcal{F}_{p_i}, \hat{\mathcal{F}}_{p_i}\right),$$

where

$$\mathcal{F} \equiv \{\mathcal{F}_{p_0}, \mathcal{F}_{p_1}, \mathcal{F}_{p_2}, \ldots\},$$
$$\hat{\mathcal{F}} \equiv \{\hat{\mathcal{F}}_{p_0}, \hat{\mathcal{F}}_{p_1}, \hat{\mathcal{F}}_{p_2}, \ldots\}. \tag{5.27}$$

We are now set up to begin the proof of parts 1, 2, and 3 of the theorem.

PROOF OF PARTS 1, 2, AND 3

The proofs of parts 1, 2, and 3 have several key ingredients that will be presented in parts.

PROOF OF THE EXISTENCE OF AN $n - (s + u)$-PARAMETER INVARIANT FAMILY OF u-DIMENSIONAL SURFACES FOR THE TIME-T MAP

We define the map

$$\mathcal{G} : S_\delta \quad \rightarrow \quad S_\delta, \tag{5.28}$$
$$\mathcal{F} \quad \mapsto \quad \mathcal{G}\mathcal{F} \equiv \hat{\mathcal{F}}, \tag{5.29}$$

where

$$\hat{\mathcal{F}}_{p_i} = G_{p_{i+1}}\left(\mathcal{F}_{p_{i+1}}\right), \quad i = 0, 1, \ldots.$$

Proposition 5.6.2 *For ϵ, δ, and η sufficiently small, $G_p : S_\delta^p \to S_\delta^{p'}$, where $\sigma_i(p) = x$ and $\sigma_j(p') = f_{ij}^0(x, 0, 0)$.*

Proof: We must show that

$$[G_p(\mathcal{F}_p)](\xi; p') \tag{5.30}$$

is Lipshitz with Lipshitz constant δ, where

$$\xi = h_{ij}^0(f_1(z; p), f_2(z; p), z).$$

From this it will also follow that the image of a graph is a graph.
    We begin with the following preliminary estimate. Let

$$\xi = h_{ij}^0(f_1(z; p), f_2(z; p), z),$$

$$\xi' = h_{ij}^0(f_1(z'; p), f_2(z'; p), z');$$

then we have

$$\| \xi - \xi' \| \geq \| h_{ij}^0(f_1(z; p), f_2(z; p), z) - h_{ij}^0(f_1(z; p), f_2(z; p), z') \|$$

$$- \| h_{ij}^0(f_1(z; p), f_2(z; p), z') - h_{ij}^0(f_1(z'; p), f_2(z; p), z') \|$$

$$- \| h_{ij}^0(f_1(z'; p), f_2(z; p), z') - h_{ij}^0(f_1(z'; p), f_2(z'; p), z') \|$$

$$\geq \left(\| (D_3 h_{ij}^0)^{-1} \|^{-1} - 2\delta \left(\| D_1 h_{ij}^0 \| + \| D_2 h_{ij}^0 \| \right) \right) \| z - z' \|, \tag{5.31}$$

where we have chosen z sufficiently close to z' (i.e., take ϵ sufficiently small) so that

$$| \mathcal{O} \left(\| z - z' \|^2 \right) | \leq \delta \left(\| D_1 h_{ij}^0 \| + \| D_2 h_{ij}^0 \| \right) \| z - z' \|.$$

Technical Detail: This type of lower bound estimate using Taylor expansions will arise repeatedly throughout the proof of this theorem, namely, an estimate of the form

$$\| \xi - \xi' \| \geq \left(\| (D_3 h_{ij}^0)^{-1} \|^{-1} - \{\text{something}\} \right) \| z - z' \| - | \mathcal{O} \left(\| z - z' \|^2 \right) |.$$

We will always handle this by choosing z and z' sufficiently small (equivalently, ϵ sufficiently small) so that

$$| \left(\| z - z' \|^2 \right) | \leq \{\text{something}\} \| z - z' \|.$$

This will result in the following estimate:

$$\| \xi - \xi' \| \geq \left(\| (D_3 h_{ij}^0)^{-1} \|^{-1} - 2\{\text{something}\} \right) \| z - z' \|.$$

Henceforth, we will pass immediately to this last estimate when the situation arises without repeatedly spelling out the details. Also, note that invertibility of $D_3 h_{ij}^0(x, 0, 0)$ follows from invariance of N^u under the linearized dynamics. Hence, for N'^u sufficiently close to N^u, and x and y sufficiently small, $D_3 h_{ij}^0(x, y, z)$ will also be invertible.

Using the definition of the graph transform, we have

$$\| \left[G\left(\mathcal{F}_p\right) \right] (\xi; p') \left[G\left(\mathcal{F}_p\right) \right] (\xi'; p') \|$$

$$= \| \left(f_{ij}^0\left(f_1(z; p), f_2(z; p), z \right) - f_{ij}^0\left(f_1(z'; p), f_2(z'; p), z' \right), \right. \qquad (5.32)$$

$$\left. g_{ij}^0\left(f_1(z; p), f_2(z; p), z \right) - g_{ij}^0\left(f_1(z'; p), f_2(z'; p), z' \right) \right) \| .$$

We estimate each component individually:

$$\| f_{ij}^0\left(f_1(z; p), f_2(z; p), z \right) - f_{ij}^0\left(f_1(z'; p), f_2(z'; p), z' \right) \|$$

$$\leq \left(\delta \left(\| D_1 f_{ij}^0 \| + \| D_2 f_{ij}^0 \| \right) + \| D_3 f_{ij}^0 \| \right) \| z - z' \|$$

$$\leq \frac{\delta \left(\| D_1 f_{ij}^0 \| + \| D_2 f_{ij}^0 \| \right) + \| D_3 f_{ij}^0 \|}{\| \left(D_3 h_{ij}^0 \right)^{-1} \|^{-1} - 2\delta \left(\| D_1 h_{ij}^0 \| + \| D_2 h_{ij}^0 \| \right)} \| \xi - \xi' \|$$

$$\leq \left(\delta \left(\| D_1 f_{ij}^0 \| + \| D_2 f_{ij}^0 \| \right) + \| D_3 f_{ij}^0 \| \right)$$

$$\times (\| \left(D_3 h_{ij}^0 \right)^{-1} \| + \mathcal{O}(\eta\delta)) \| \xi - \xi' \| . \qquad (5.33)$$

It follows from the GR-I estimates that by taking η and δ sufficiently small we have

$$\left(\delta \left(\| D_1 f_{ij}^0 \| + \| D_2 f_{ij}^0 \| \right) + \| D_3 f_{ij}^0 \| \right) \left(\| \left(D_3 h_{ij}^0 \right)^{-1} \| + \mathcal{O}(\eta\delta) \right) < \delta.$$

Similarly, we have

$$\| g_{ij}^0\left(f_1(z; p), f_2(z; p), z \right) - g_{ij}^0\left(f_1(z'; p), f_2(z'; p), z' \right) \|$$

$$\leq \left(\delta \left(\| D_1 g_{ij}^0 \| + \| D_2 g_{ij}^0 \| \right) + \| D_3 g_{ij}^0 \| \right) \| z - z' \|$$

$$\leq \frac{\delta \left(\| D_1 g_{ij}^0 \| + \| D_2 g_{ij}^0 \| \right) + \| D_3 g_{ij}^0 \|}{\| \left(D_3 h_{ij}^0 \right)^{-1} \|^{-1} - 2\delta \left(\| D_1 h_{ij}^0 \| + \| D_2 h_{ij}^0 \| \right)} \| \xi - \xi' \|$$

$$\leq \left(\delta \left(\| D_1 g_{ij}^0 \| + \| D_2 g_{ij}^0 \| \right) + \| D_3 g_{ij}^0 \| \right)$$

$$\times (\| \left(D_3 h_{ij}^0 \right)^{-1} \| + \mathcal{O}(\eta\delta)) \| \xi - \xi' \| . \qquad (5.34)$$

Again, it follows from the GR-I estimates that by taking η and δ sufficiently small we have

$$\left(\delta\left(\parallel D_1 g_{ij}^0 \parallel + \parallel D_2 g_{ij}^0 \parallel\right) + \parallel D_3 g_{ij}^0 \parallel\right)\left(\parallel \left(D_3 h_{ij}^0\right)^{-1} \parallel + \mathcal{O}(\eta\delta)\right) < \delta.$$

Thus, the proposition is proved.   □

> **Technical Detail:** Showing that the graph transform is well defined in the case of foliations of the unstable manifold is much easier than in the cases for the persisting overflowing invariant manifold and for the unstable manifold. The major difference is that the latter two manifolds are global, i.e., they are defined over all the chart domains of $\bar{M}$, whereas a given fiber is defined in one chart domain.

Proposition 5.6.3 *For ϵ, δ, and η sufficiently small, G_p is a contraction map of S_δ^p into $S_\delta^{p'}$, where $\sigma_i(p) = x$ and $\sigma_j(p') = f_{ij}^0(x, 0, 0)$.*

Proof: Choose f_1, f_2, $\hat{f}_1$, $\hat{f}_2$, z, z' such that

$$\xi = h_{ij}^0\left(f_1(z;p), f_2(z;p), z\right) = h_{ij}^0(\hat{f}_1(z';p), \hat{f}_2(z';p), z'). \tag{5.35}$$

First we need some preliminary estimates.

Expansion Estimates

Consider the following two expressions:

$$h_{ij}^0(\hat{f}_1(z';p), \hat{f}_2(z';p), z) - h_{ij}^0(\hat{f}_1(z';p), \hat{f}_2(z';p), z'), \tag{5.36}$$

$$h_{ij}^0(\hat{f}_1(z';p), \hat{f}_2(z';p), z) - h_{ij}^0\left(f_1(z;p), f_2(z;p), z\right). \tag{5.37}$$

Note that by (5.35), these two expressions are equal. We want to obtain a lower bound for the first expression and an upper bound for the second expression. We have

$$\parallel h_{ij}^0(\hat{f}_1(z';p), \hat{f}_2(z';p), z) - h_{ij}^0(\hat{f}_1(z';p), \hat{f}_2(z';p), z') \parallel$$

$$\geq \parallel \left(D_3 h_{ij}^0\right)^{-1} \parallel^{-1} \parallel z - z' \parallel - |\mathcal{O}(\parallel z - z' \parallel^2)| \tag{5.38}$$

and

$$\parallel h_{ij}^0\left(\hat{f}_1(z';p), \hat{f}_2(z';p), z\right) - h_{ij}^0\left(f_1(z;p), f_2(z;p), z\right) \parallel$$

$$\leq \parallel h_{ij}^0\left(\hat{f}_1(z';p), \hat{f}_2(z';p), z\right) - h_{ij}^0\left(\hat{f}_1(z;p), f_2(z;p), z\right) \parallel$$

$$+ \parallel h_{ij}^0 \left(\hat{f}_1(z;p), f_2(z';p), z \right) - h_{ij}^0 \left(f_1(z;p), f_2(z;p), z \right) \parallel$$

$$\leq \parallel D_1 h_{ij}^0 \parallel \delta \parallel z - z' \parallel + \parallel D_2 h_{ij}^0 \parallel d_2^p \left(f_2, \hat{f}_2 \right) \parallel z' \parallel$$

$$+ \parallel D_1 h_{ij}^0 \parallel d_1^p \left(f_1, \hat{f}_1 \right) \parallel z \parallel + \parallel D_2 h_{ij}^0 \parallel \delta \parallel z - z' \parallel . \quad (5.39)$$

Combining these two estimates gives:

$$\parallel z - z' \parallel \leq \frac{\parallel D_1 h_{ij}^0 \parallel d_1^p(f_1, \hat{f}_1) \parallel z \parallel + \parallel D_2 h_{ij}^0 \parallel d_2^p(f_2, \hat{f}_2) \parallel z' \parallel}{\parallel \left(D_3 h_{ij}^0 \right)^{-1} \parallel^{-1} - 2\delta \left(\parallel D_1 h_{ij}^0 \parallel + \parallel D_2 h_{ij}^0 \parallel \right)} . \quad (5.40)$$

Next we want to estimate $\parallel z \parallel$ and $\parallel z' \parallel$ in terms of ξ.

$$\parallel \xi \parallel = \parallel h_{ij}^0 \left(f_1(z;p), f_2(z;p), z \right) \parallel$$

$$= \parallel h_{ij}^0 \left(f_1(z;p), f_2(z;p), z \right) - h_{ij}^0 \left(f_1(z;p), 0, 0 \right) \parallel$$

$$\geq \parallel h_{ij}^0 \left(f_1(z;p), f_2(z;p), z \right) - h_{ij}^0 \left(f_1(z;p), f_2(z;p), 0 \right) \parallel$$

$$- \parallel h_{ij}^0 \left(f_1(z;p), f_2(z;p), 0 \right) - h_{ij}^0 \left(f_1(z;p), 0, 0 \right) \parallel$$

$$\geq \left(\parallel \left(D_3 h_{ij}^0 \right)^{-1} \parallel^{-1} - 2\delta \parallel D_2 h_{ij}^0 \parallel \right) \parallel z \parallel . \quad (5.41)$$

Similarly, using (5.35), we have

$$\parallel \xi \parallel = \parallel h_{ij}^0 (\hat{f}_1(z';p), \hat{f}_2(z';p), z') \parallel$$

$$= \parallel h_{ij}^0 (\hat{f}_1(z';p), \hat{f}_2(z';p), z') - h_{ij}^0 (\hat{f}_1(z';p), 0, 0) \parallel$$

$$\geq \parallel h_{ij}^0 (\hat{f}_1(z';p), \hat{f}_2(z';p), z') - h_{ij}^0 (\hat{f}_1(z';p), \hat{f}_2(z';p), 0) \parallel$$

$$- \parallel h_{ij}^0 (\hat{f}_1(z';p), \hat{f}_2(z';p), 0) - h_{ij}^0 (\hat{f}_1(z';p), 0, 0) \parallel$$

$$\geq \left(\parallel \left(D_3 h_{ij}^0 \right)^{-1} \parallel^{-1} - 2\delta \parallel D_2 h_{ij}^0 \parallel \right) \parallel z' \parallel . \quad (5.42)$$

Main Estimate

$$\| \left[G_p\left(\mathcal{F}_p\right)\right] (\xi; p') \left[G_p(\hat{\mathcal{F}}_p)\right] (\xi; p') \|$$

$$=\| \Big(f_{ij}^0 \left(f_1(z;p), f_2(z;p), z\right) - f_{ij}^0(\hat{f}_1(z';p), \hat{f}_2(z';p), z'),$$

$$g_{ij}^0 \left(f_1(z;p), f_2(z;p), z\right) - g_{ij}^0(\hat{f}_1(z';p), \hat{f}_2(z';p), z') \Big) \| . \tag{5.43}$$

We estimate each component separately.

$$\| f_{ij}^0 \left(f_1(z;p), f_2(z;p), z\right) - f_{ij}^0(\hat{f}_1(z';p), \hat{f}_2(z';p), z') \|$$

$$\leq \| f_{ij}^0 \left(f_1(z;p), f_2(z;p), z\right) - f_{ij}^0 \left(f_1(z';p), f_2(z;p), z\right) \|$$

$$+ \| f_{ij}^0 \left(f_1(z';p), f_2(z;p), z\right) - f_{ij}^0(\hat{f}_1(z';p), f_2(z;p), z) \|$$

$$+ \| f_{ij}^0(\hat{f}_1(z';p), f_2(z;p), z) - f_{ij}^0(\hat{f}_1(z';p), f_2(z';p), z) \|$$

$$+ \| f_{ij}^0(\hat{f}_1(z';p), f_2(z';p), z) - f_{ij}^0(\hat{f}_1(z';p), \hat{f}_2(z';p), z) \|$$

$$+ \| f_{ij}^0(\hat{f}_1(z';p), \hat{f}_2(z';p), z) - f_{ij}^0(\hat{f}_1(z';p), \hat{f}_2(z';p), z') \|$$

$$\leq \left(\delta \left(\| D_1 f_{ij}^0 \| + \| D_2 f_{ij}^0 \|\right) + \| D_3 f_{ij}^0 \|\right) \| z - z' \|$$

$$+ \| D_1 f_{ij}^0 \| d_1^p(f_1, \hat{f}_1) \| z' \| + \| D_2 f_{ij}^0 \| d_2^p(f_2, \hat{f}_2) \| z' \|$$

$$\leq \frac{\left(\delta\left(\|D_1 f_{ij}^0\|+\|D_2 f_{ij}^0\|\right)+\|D_3 f_{ij}^0\|\right)\left(\|D_1 h_{ij}^0\|d_1^p\left(f_1, \hat{f}_1\right)+\|D_2 h_{ij}^0\|d_2^p\left(f_2, \hat{f}_2\right)\right)}{\left(\|\left(D_3 h_{ij}^0\right)^{-1}\|^{-1}-2\delta\left(\|D_1 h_{ij}^0\|+\|D_2 h_{ij}^0\|\right)\right)\left(\|\left(D_3 h_{ij}^0\right)^{-1}\|^{-1}-2\delta\|D_2 h_{ij}^0\|\right)} \| \xi \|$$

$$+\frac{\left(\|D_1 f_{ij}^0\|d_1^p\left(f_1, \hat{f}_1\right)+\|D_2 f_{ij}^0\|d_2^p\left(f_2, \hat{f}_2\right)\right)}{\left(\|\left(D_3 h_{ij}^0\right)^{-1}\|^{-1}-2\delta\|D_2 h_{ij}^0\|\right)} \| \xi \|$$

$$\leq \left(\delta \| \left(D_3 h_{ij}^0\right)^{-1} \| \| D_1 f_{ij}^0 \| + \mathcal{O}(\eta\delta)\right) d_1^p(f_1, \hat{f}_1) \| \xi \|$$

$$+\left(\| \left(D_3 h_{ij}^0\right)^{-1} \| \| D_2 f_{ij}^0 \| + \mathcal{O}(\eta\delta)\right) d_2^p(f_2, \hat{f}_2) \| \xi \|, \tag{5.44}$$

$$\| g_{ij}^0 \left(f_1(z;p), f_2(z;p), z \right) - g_{ij}^0 (\hat{f}_1(z';p), \hat{f}_2(z';p), z') \|$$

$$\leq \| g_{ij}^0 \left(f_1(z;p), f_2(z;p), z \right) - g_{ij}^0 \left(f_1(z';p), f_2(z;p), z \right) \|$$

$$+ \| g_{ij}^0 (f_1(z';p), f_2(z;p), z) - g_{ij}^0 (\hat{f}_1(z';p), f_2(z;p), z) \|$$

$$+ \| g_{ij}^0 (\hat{f}_1(z';p), f_2(z;p), z) - g_{ij}^0 (\hat{f}_1(z';p), f_2(z';p), z) \|$$

$$+ \| g_{ij}^0 (\hat{f}_1(z';p), f_2(z';p), z) - g_{ij}^0 (\hat{f}_1(z';p), \hat{f}_2(z';p), z) \|$$

$$+ \| g_{ij}^0 (\hat{f}_1(z';p), \hat{f}_2(z';p), z) - g_{ij}^0 (\hat{f}_1(z';p), \hat{f}_2(z';p), z') \|$$

$$\leq \left(\delta \left(\| D_1 g_{ij}^0 \| + \| D_2 g_{ij}^0 \| \right) + \| D_3 g_{ij}^0 \| \right) \| z - z' \|$$

$$+ \| D_1 g_{ij}^0 \| d_1^p(f_1, \hat{f}_1) \| z' \| + \| D_2 g_{ij}^0 \| d_2^p(f_2, \hat{f}_2) \| z' \|$$

$$\leq \frac{\left(\delta \left(\| D_1 g_{ij}^0 \| + \| D_2 g_{ij}^0 \| \right) + \| D_3 g_{ij}^0 \| \right) \left(\| D_1 h_{ij}^0 \| d_1^p \left(f_1, \hat{f}_1 \right) + \| D_2 h_{ij}^0 \| d_2^p \left(f_2, \hat{f}_2 \right) \right)}{\left(\| (D_3 h_{ij}^0)^{-1} \|^{-1} - 2\delta \left(\| D_1 h_{ij}^0 \| + \| D_2 h_{ij}^0 \| \right) \right) \left(\| (D_3 h_{ij}^0)^{-1} \|^{-1} - 2\delta \| D_2 h_{ij}^0 \| \right)} \| \xi \|$$

$$+ \frac{\left(\| D_1 g_{ij}^0 \| d_1^p \left(f_1, \hat{f}_1 \right) + \| D_2 g_{ij}^0 \| d_2^p \left(f_2, \hat{f}_2 \right) \right)}{\left(\| (D_3 h_{ij}^0)^{-1} \|^{-1} - 2\delta \| D_2 h_{ij}^0 \| \right)} \| \xi \|$$

$$\leq (\mathcal{O}(\eta\delta) + \mathcal{O}(\eta)) d_1^p(f_1, \hat{f}_1) \| \xi \|$$

$$+ (\| (D_3 h_{ij}^0)^{-1} \| \| D_2 g_{ij}^0 \| + \mathcal{O}(\eta\delta)) d_2^p(f_2, \hat{f}_2) \| \xi \| . \tag{5.45}$$

It thus follows from (5.43), (5.44), and (5.45) that

$$d^{p'} \left(G_p(\mathcal{F}_p), G_p(\hat{\mathcal{F}}_p) \right) \leq C d^p(\mathcal{F}_p, \hat{\mathcal{F}}_p), \tag{5.46}$$

where

$$C \equiv \max \left\{ \| (D_3 h_{ij}^0)^{-1} \| \| D_1 f_{ij}^0 \| + \mathcal{O}(\eta\delta) , \right.$$

$$\| (D_3 h_{ij}^0)^{-1} \| \| D_2 f_{ij}^0 \| + \mathcal{O}(\eta\delta), \mathcal{O}(\eta\delta) + \mathcal{O}(\eta),$$

$$\left. \| (D_3 h_{ij}^0)^{-1} \| \| D_2 g_{ij}^0 \| + \mathcal{O}(\eta\delta) \right\}.$$

Hence, for ϵ, η and δ sufficiently small, $C < 1$. Thus, appealing to Proposition 5.5.1 completes the proof. $\square$

PROOF THAT THE INVARIANT FAMILY FOR THE TIME-T MAP IS C^r

The proof goes the same as the proof of the differentiability of the persisting overflowing invariant manifold and the unstable manifold. The individual surfaces in the invariant family satisfy

$$f_1(\xi; p') = f_{ij}^0 (f_1(z; p), f_2(z; p), z),$$

$$f_2(\xi; p') = g_{ij}^0 (f_1(z; p), f_2(z; p), z),$$

(5.47)

where

$$\xi = h_{ij}^0 (f_1(z; p), f_2(z; p), z),$$

$$\sigma_j(p') = f_{ij}^0(x, 0, 0), \quad f_1(0; p) = x, \quad f_2(0; p) = 0.$$

Formally differentiating these equations gives

$$Df_1(\xi; p') = \left[D_1 f_{ij}^0 D f_1 + D_2 f_{ij}^0 D f_2 + D_3 f_{ij}^0 \right]$$

$$\times \left[D_1 h_{ij}^0 D f_1 + D_2 h_{ij}^0 D f_2 + D_3 h_{ij}^0 \right]^{-1},$$

$$Df_2(\xi; p') = \left[D_1 g_{ij}^0 D f_1 + D_2 g_{ij}^0 D f_2 + D_3 g_{ij}^0 \right]$$

(5.48)

$$\times \left[D_1 h_{ij}^0 D f_1 + D_2 h_{ij}^0 D f_2 + D_3 h_{ij}^0 \right]^{-1},$$

where on the right-hand side of these equations the coordinate expressions for the derivatives of the components of the time-T map are evaluated on $(f_1(z; p), f_2(z; p), z)$ and the derivatives of f_1 and f_2 are evaluated at z, p. The procedure now is as follows:

1. Set up an iteration scheme and show that these equations have a solution.

2. Show that the solution thus obtained is actually the derivative (i.e., it satisfies the definition of the derivative).

3. Successively differentiate and show the existence of higher order derivatives, up to r.

The proof goes exactly like that for the persisting overflowing invariant manifold or the unstable manifold. The fact that the iteration procedure defines a contraction map follows from the following two quantities:

$$\| D_1 f_{ij}^0 \| \, \| (D_3 h_{ij}^0)^{-1} \| < \min \left\{ \frac{\rho^T}{8}, \frac{\rho^T}{8K} \right\} < \min \left\{ \frac{1}{8}, \frac{1}{8K} \right\},$$

(5.49)

$$\| D_2 g_{ij}^0 \| \, \| (D_3 h_{ij}^0)^{-1} \| < \min \left\{ \frac{\rho^T}{8}, \frac{\rho^T}{8K} \right\} < \min \left\{ \frac{1}{8}, \frac{1}{8K} \right\}.$$

The remaining terms in the formal equations for the derivative are small by the GR-I estimates. The fact that we can only obtain r derivatives follows from the fact that the time-T map is only C^r by assumption.

PROOF THAT THE ELEMENTS OF THE INVARIANT FAMILY OF THE
TIME-T MAP ARE TANGENT TO FIBERS OF N'^u

The proof of this goes exactly as the proof of the analogous result for the proof of the unstable manifold.

PROOF THAT THE INVARIANT FAMILY FOR THE TIME-T MAP IS AN
INVARIANT FAMILY FOR THE TIME-T MAP, FOR ARBITRARY $t < 0$

The proof of this is the same as that for the persistence of overflowing invariant manifolds. Recall that the main idea was that for t sufficiently small, the image of a Lipshitz graph is still a Lipshitz graph. One then uses commutativity of ϕ_t with ϕ_T, along with invariance under the time-T flow map and repeats the argument; see Corollary 3.3.4.

We next prove parts 4 and 5 dealing with estimates on the rates and nature of the approach of points to $\bar{M}$ along the fibers asymptotically as $t \to -\infty$. However, first we must establish some notation.

Notation

Let $p \in M$ and let us consider the sequence

$$\{p \equiv p_0, p_1, p_2, \ldots\},$$

where

$$p_i = \phi_{-T}(p_{i-1}), \quad i \geq 1.$$

In local coordinates, we have

$$\sigma_i(p_k) \equiv x_k$$

and

$$\sigma_j(p_k) = f_{ij}^0(x_{k+1}, 0, 0).$$

In local coordinates, we denote the unstable fiber through the basepoint p_k as

$$(f_1(z_k; p_k), f_2(z_k; p_k), z_k) \equiv (x_k, y_k, z_k),$$

where $\| z_k \| < \epsilon$. Two different points on the *same* fiber will be distinguished by primes, i.e.,

$$(x_k, y_k, z_k) \qquad \text{and} \qquad (x'_k, y'_k, z'_k).$$

Two different fibers, i.e., fibers through different basepoints, will be distinguished with "hats" as follows:

$$(f_1(z_k; p_k), f_2(z_k; p_k), z_k) \equiv (x_k, y_k, z_k)$$

and

$$(f_1(z_k; \hat{p}_k), f_2(z_k; \hat{p}_k), z_k) \equiv (\hat{x}_k, \hat{y}_k, z_k).$$

We are now ready to begin the proofs of parts 4, 5, and 6 of the theorem.

PROOF OF PART 5

Let (x_k, y_k, z_k) denote the local coordinate representation for $\phi_{-kT}(q)$, and let (x'_k, y'_k, z'_k) denote the local coordinate representation for $\phi_{-kT}(q')$, where $q, q' \in f^u(p)$. From (5.31) and the GR-I estimates we have

$$\begin{aligned}
\| z_k - z'_k \| &\leq \ (\| (D_3 h^0_{ij})^{-1} \| + \mathcal{O}(\eta\delta)) \ \| z_{k-1} - z'_{k-1} \| \\
&\leq \ C \ \| z_{k-1} - z'_{k-1} \|,
\end{aligned} \tag{5.50}$$

for η and δ sufficiently small, where

$$C \equiv \min\left\{ \frac{a^T}{4}, \frac{a^T}{4\mathcal{K}} \right\},$$

with

$$\mathcal{K} \equiv \| D_2 f^0_{ij} \|.$$

Let us introduce the notation

$$d_k = \| z_k - z'_k \|.$$

Then (5.50) can be rewritten as

$$d_k \leq C d_{k-1}.$$

We define

$$d_t \equiv \| z(-t) - z'(-t) \|,$$

and for any fixed integer $k \geq 1$ let

$$M_k \equiv \max_{(k-1)T \leq t \leq kT} d_t.$$

M_k exists since d_t is a continuous function and we are considering a compact domain. Since we are free to choose the origin of time, we have

$$M_{k+1} \leq C M_k. \tag{5.51}$$

Also, we can find constants $C' > 0$ and $\lambda_u > 0$ such that

$$\max_{0 \leq t \leq T} d_t < C'e^{-\lambda_u t}, \quad t \in [0, T].$$

$$(5.52)$$

We want to argue that

$$d_t \leq C_0 e^{-\lambda_u t}, \qquad \forall t > 0.$$

$$(5.53)$$

We will do this by showing that

$$\max_{(k-1)T \leq t \leq kT} d_t \leq C_0 e^{-\lambda_u t} \quad \text{for all integers } k \geq 1,$$

from which (5.53) follows.

The argument proceeds as follows. Choose $\lambda_u = -\log a$. Then the constant C' from (5.52) must satisfy

$$C'a^T > \max_{0 \leq t \leq T} d_t,$$

from which it follows that

$$C' > \frac{1}{a^T} \max_{0 \leq t \leq T} d_t = \frac{1}{a^T} M_1.$$

$$(5.54)$$

Now from (5.51) and the definition of C for any integer $k \geq 1$ we have

$$\max_{(k-1)T \leq t \leq kT} d_t \leq \tilde{C} a^{(k-1)T} M_1,$$

$$(5.55)$$

where

$$\tilde{C} = \min \left\{ \left(\frac{1}{4}\right)^{k-1}, \left(\frac{1}{4\mathcal{K}}\right)^{k-1} \right\}.$$

With our choice for λ_u we have the following inequalities:

$$\min_{(k-1)T \leq t \leq kT} C'e^{-\lambda_u t} \geq \frac{1}{a^T} \max_{0 \leq t \leq T} d_t a^{kT} \quad \text{(using (5.54))}$$

$$\geq \max_{0 \leq t \leq T} d_t a^{(k-1)T} = M_1 a^{(k-1)T}$$

$$\geq \frac{1}{\tilde{C}} \max_{(k-1)T \leq t \leq kT} d_t \quad \text{(using (5.55)). (5.56)}$$

From these last inequalities it follows that

$$\max_{(k-1)T \leq t \leq kT} d_t \leq \tilde{C} C' e^{-\lambda_u t}, \qquad \forall t > 0,$$

and we take $C_0 = \tilde{C} C'$.

Thus, we have shown that

$$\| z(-t) - z'(-t) \| \le C e^{-\lambda_u t}, \qquad t \to \infty, \qquad (5.57)$$

where, for simplicity of notation, we have omitted denoting the local coordinate changes that may occur as $t \to \infty$. Furthermore, we can write

$$\| y(-t) - y'(-t) \| = \| f_2(z(-t); p(-t)) - f_2'(z'(-t); p'(-t)) \|$$

$$\le \delta \| z(-t) - z'(-t) \|,$$

$$\| x(-t) - x'(-t) \| = \| f_1(z(-t); p(-t)) - f_1'(z'(-t); p'(-t)) \|$$

$$\le \delta \| z(-t) - z'(-t) \|, \qquad (5.58)$$

where, again, for simplicity of notation, we have omitted denoting the local coordinate changes that may occur as $t \to \infty$. Thus, using (5.57) and (5.58) gives

$$\| \phi_{-t}(q) - \phi_{-t}(q') \| < \hat{C} \left(\| x(-t) - x'(-t) \| + \| y(-t) - y'(-t) \| \right.$$

$$+ \| z(-t) - z'(-t) \|) < \hat{C} C_0 (1 + 2\delta) e^{-\lambda_u t}.$$

$$(5.59)$$

By setting $C_u = \hat{C} C_0 (1 + 2\delta)$ and $q' = p$ we complete the proof of part 5.

PROOF OF PART 6

It is enough to show that

$$\frac{\| \phi_{-t}(q) - \phi_{-t}(p) \|}{\| \phi_{-t}(p') - \phi_{-t}(p) \|} \to 0 \quad \text{as} \quad t \to \infty$$

and

$$\frac{\| \phi_{-t}(q') - \phi_{-t}(p') \|}{\| \phi_{-t}(p') - \phi_{-t}(p) \|} \to 0 \quad \text{as} \quad t \to \infty.$$

This can be seen by noting that

$$\frac{\| \phi_{-t}(q) - \phi_{-t}(p) \|}{\| \phi_{-t}(q') - \phi_{-t}(p) \|} = \frac{\| \phi_{-t}(q) - \phi_{-t}(p) \|}{\| \phi_{-t}(q') - \phi_{-t}(p') + \phi_{-t}(p') - \phi_{-t}(p) \|}$$

$$\le \frac{\| \phi_{-t}(q) - \phi_{-t}(p) \|}{\left| \| \phi_{-t}(q') - \phi_{-t}(p') \| - \| \phi_{-t}(p') - \phi_{-t}(p) \| \right|}$$

$$= \frac{1}{\left| \frac{\| \phi_{-t}(q') - \phi_{-t}(p') \|}{\| \phi_{-t}(q) - \phi_{-t}(p) \|} - \frac{\| \phi_{-t}(p') - \phi_{-t}(p) \|}{\| \phi_{-t}(q) - \phi_{-t}(p) \|} \right|}$$

$$= \frac{1}{\left| \frac{\| \phi_{-t}(q') - \phi_{-t}(p') \|}{\| \phi_{-t}(p') - \phi_{-t}(p) \|} \frac{\| \phi_{-t}(p') - \phi_{-t}(p) \|}{\| \phi_{-t}(q) - \phi_{-t}(p) \|} - \frac{\| \phi_{-t}(p') - \phi_{-t}(p) \|}{\| \phi_{-t}(q) - \phi_{-t}(p) \|} \right|}$$

$$= \frac{1}{\left| \frac{\| \phi_{-t}(p') - \phi_{-t}(p) \|}{\| \phi_{-t}(q) - \phi_{-t}(p) \|} \left(\frac{\| \phi_{-t}(q') - \phi_{-t}(p') \|}{\| \phi_{-t}(p') - \phi_{-t}(p) \|} - 1 \right) \right|} .$$

From this expression we can conclude that if

$$\frac{\| \phi_{-t}(p') - \phi_{-t}(p) \|}{\| \phi_{-t}(q) - \phi_{-t}(p) \|} \to \infty \quad \text{as} \quad t \to \infty$$

and

$$\frac{\| \phi_{-t}(q') - \phi_{-t}(p') \|}{\| \phi_{-t}(p') - \phi_{-t}(p) \|} \to 0 \quad \text{as} \quad t \to \infty,$$

then

$$\frac{\| \phi_{-t}(q) - \phi_{-t}(p) \|}{\| \phi_{-t}(q') - \phi_{-t}(p) \|} \to 0 \quad \text{as} \quad t \to \infty.$$

The local coordinate expression for

$$\frac{\| \phi_{-t}(q) - \phi_{-t}(p) \|}{\| \phi_{-t}(p') - \phi_{-t}(p) \|} \to 0 \quad \text{as} \quad t \to \infty$$

is

$$\frac{\| (f_1(z_k; p_k), f_2(z_k; p_k), z_k) - (f_1(0; p_k), 0, 0) \|}{\| \left(\hat{f}_1(0; \hat{p}_k), 0, 0 \right) - (f_1(0; p_k), 0, 0)) \|} \tag{5.60}$$

or

$$\frac{\| (x_k - x'_k, y_k, z_k) \|}{\| x_k - \hat{x}_k \|} . \tag{5.61}$$

We first first recall some estimates in our new notation. From (5.33) and (5.34) we have

$$\begin{aligned}
\| x_k - x'_k \| &= \| f_1(z_k; p_k) - f_1(z'_k; p_k) \| \le \delta \| z_k - z'_k \|, \\
\| y_k - y'_k \| &= \| f_2(z_k; p_k) - f_2(z'_k; p_k) \| \le \delta \| z_k - z'_k \| . \tag{5.62}
\end{aligned}$$

These estimates are for two points *on the same fiber*. The following estimate is for the basepoints *on two different fibers*.

Recall that for two basepoints in local coordinates we have

$$x_{k-1} = f_{ij}^0(x_k, 0, 0),$$
$$\hat{x}_{k-1} = f_{ij}^0(\hat{x}_k, 0, 0),$$

Subtracting and Taylor expanding gives

$$\| x_{k-1} - \hat{x}_{k-1} \| \leq \left(\| \left(D_1 f_{ij}^0 \right) \| + \eta \right) \| x_k - \hat{x}_k \| \qquad (5.63)$$

or

$$\| x_k - \hat{x}_k \| \geq \left(\| \left(D_1 f_{ij}^0 \right) \| + \eta \right)^{-1} \| x_{k-1} - \hat{x}_{k-1} \| \qquad (5.64)$$

where we are assuming that x_k and $\hat{x}_k$ are close enough so that

$$\mathcal{O} \left(\| x_k - \hat{x}_k \|^2 \right) \leq \eta \| x_k - \hat{x}_k \| .$$

Also, recall from (5.31) and the GR-I estimates that

$$\| z_k - z_k' \| \leq \left(\| \left(D_3 h_{ij}^0 \right)^{-1} \| + \mathcal{O}(\eta \delta) \right) \| z_{k-1} - z_{k-1}' \| .$$

Using these estimates, we estimate (5.61) as follows:

$$
\begin{aligned}
\frac{\| (x_k - x_k', y_k, z_k) \|}{\| x_k - \hat{x}_k \|} &\leq \frac{\delta \| z_k - z_k' \| + \delta \| z_k - z_k' \| + \| z_k - z_k' \|}{\left(\| D_1 f_{ij}^0 \| + \eta \right)^{-1} \| x_{k-1} - \hat{x}_{k-1} \|} \\
&\leq \frac{(2\delta + 1) \left(\| \left(D_3 h_{ij}^0 \right)^{-1} \| + \mathcal{O}(\eta \delta) \right) \| z_{k-1} \|}{\left(\| D_1 f_{ij}^0 \| + \eta \right)^{-1} \| x_{k-1} - \hat{x}_{k-1} \|} \\
&= \left(\| \left(D_3 h_{ij}^0 \right)^{-1} \| \| D_1 f_{ij}^0 \| + \mathcal{O}(\eta) + \mathcal{O}(\delta) \right) \frac{\| z_{k-1} \|}{\| x_{k-1} - \hat{x}_{k-1} \|} .
\end{aligned}
$$
$$(5.65)$$

Using the inequality

$$\| (x_{k-1} - x_{k-1}', y_{k-1}, z_{k-1}) \| \geq \| z_{k-1} \|,$$

and (5.65), (5.61) becomes

$$\frac{\| (x_k - x_k', y_k, z_k) \|}{\| x_k - \hat{x}_k \|}$$

$$\leq \left(\| \left(D_3 h_{ij}^0 \right)^{-1} \| \| D_1 f_{ij}^0 \| + \mathcal{O}(\eta) + \mathcal{O}(\delta) \right) \frac{\| (x_{k-1} - x_{k-1}', y_{k-1}, z_{k-1}) \|}{\| x_{k-1} - \hat{x}_{k-1} \|} .$$
$$(5.66)$$

Using the GR-I estimates and (5.66) we obtain

$$\lim_{k \to \infty} \frac{\| (x_k - x_k', y_k, z_k) \|}{\| x_k - \hat{x}_k \|} = 0$$

or, equivalently,

$$\lim_{t \to \infty} \frac{\| \phi_{-t}(q) - \phi_{-t}(p) \|}{\| \phi_{-t}(p') - \phi_{-t}(p) \|} = 0.$$

The calculation for

$$\frac{\| \phi_{-t}(q') - \phi_{-t}(p') \|}{\| \phi_{-t}(p') - \phi_{-t}(p) \|} \to 0 \quad \text{as} \quad t \to \infty$$

proceeds in exactly the same manner.

PROOF OF PART 7

The proof is by contradiction. We assume that $f^u(p_1) \cap f^u(p_2) \ni q$ but $f^u(p_1) \neq f^u(p_2)$. Then there exists $q_2 \in f^u(p_2)$ but $q_2 \notin f^u(p_1)$, and $q_1 \in f^u(p_1)$ but $q_1 \notin f^u(p_2)$. By part 5 we can write

$$
\begin{aligned}
0 &= \lim_{t \to \infty} \frac{\| \phi_{-t}(q) - \phi_{-t}(p_2) \|}{\| \phi_{-t}(q_1) - \phi_{-t}(p_2) \|} \\
&\geq \lim_{t \to \infty} \frac{\big| \, \| \phi_{-t}(q) - \phi_{-t}(q_1) \| - \| \phi_{-t}(q_1) - \phi_{-t}(p_2) \| \, \big|}{\| \phi_{-t}(q_1) - \phi_{-t}(p_2) \|} \\
&= \lim_{t \to \infty} \left| \frac{\| \phi_{-t}(q) - \phi_{-t}(q_1) \|}{\| \phi_{-t}(q_1) - \phi_{-t}(p_2) \|} - 1 \right|.
\end{aligned}
$$

This immediately gives

$$\lim_{t \to \infty} \frac{\| \phi_{-t}(q) - \phi_{-t}(q_1) \|}{\| \phi_{-t}(q_1) - \phi_{-t}(p_2) \|} = 1. \tag{5.67}$$

Similarly, using part 6, we have

$$
\begin{aligned}
0 &= \lim_{t \to \infty} \frac{\| \phi_{-t}(q) - \phi_{-t}(p_1) \|}{\| \phi_{-t}(q_2) - \phi_{-t}(p_1) \|} \\
&\geq \lim_{t \to \infty} \frac{\big| \, \| \phi_{-t}(q) - \phi_{-t}(q_1) \| - \| \phi_{-t}(q_1) - \phi_{-t}(p_1) \| \, \big|}{\| \phi_{-t}(q_2) - \phi_{-t}(p_1) \|} \\
&= \lim_{t \to \infty} \left| \frac{\| \phi_{-t}(q) - \phi_{-t}(q_1) \|}{\| \phi_{-t}(q_2) - \phi_{-t}(p_1) \|} - \frac{\| \phi_{-t}(q_1) - \phi_{-t}(p_1) \|}{\| \phi_{-t}(q_2) - \phi_{-t}(p_1) \|} \right| \\
&= \lim_{t \to \infty} \frac{\| \phi_{-t}(q) - \phi_{-t}(q_1) \|}{\| \phi_{-t}(q_2) - \phi_{-t}(p_1) \|},
\end{aligned}
$$

which yields

$$\lim_{t \to \infty} \frac{\| \phi_{-t}(q) - \phi_{-t}(q_1) \|}{\| \phi_{-t}(q_2) - \phi_{-t}(p_1) \|} = 0. \tag{5.68}$$

Combining (5.67) and (5.68) gives

$$\lim_{t \to \infty} \frac{\| \phi_{-t}(q_1) - \phi_{-t}(p_2) \|}{\| \phi_{-t}(q_2) - \phi_{-t}(p_1) \|} = 0. \tag{5.69}$$

Since neither p_1 or p_2 are distinguished, we can interchange the indices in (5.69) to obtain

$$\lim_{t \to \infty} \frac{\| \phi_{-t}(q_2) - \phi_{-t}(p_1) \|}{\| \phi_{-t}(q_1) - \phi_{-t}(p_2) \|} = 0. \tag{5.70}$$

Now note that the expressions on the left-hand sides of (5.69) and (5.70) are reciprocals of each other. Hence, we have established a contradiction.

PROOF THAT THE FIBERS ARE C^r WITH RESPECT TO THE BASEPOINTS

We follow Fenichel [1977]. However, we are able to improve his result slightly in the context of normally hyperbolic invariant manifolds by gaining an extra derivative. If we show that the set

$$\Sigma = \{(p_1, p_2) \,|\, p_1 \in M, \; p_2 \in f^u(p_1)\}$$

is a C^r submanifold of $M \times \mathbb{R}^n$, then it follows that the fibers are C^r with respect to the basepoint.

Let $U \subset \mathbb{R}^n$ be a neighborhood of M in which the family $\mathcal{F}^u$ has been constructed. Note that Σ is the image of $h^u(N'^u) \cap U$ under the map

$$\mathcal{W} : h^u(N'^u) \cap U \quad \to \quad M \times \mathbb{R}^n, \tag{5.71}$$

$$(x, z) \quad \mapsto \quad (x, f_1(z; x), z, f_2(z; x)), \tag{5.72}$$

where we identified the coordinate x of the point p with p itself (see the definitions of f_1 and f_2). At this point we do not know about the smoothness properties of $f_1(z; x)$ and $f_2(z; x)$ with respect to x. However, by construction, we have

$$\begin{aligned} D_x f_1(0; x) &= Id_{n-(s+u)}, \\ D_x f_2(0; x) &= 0, \end{aligned} \tag{5.73}$$

where $Id_{n-(s+u)}$ denotes the $n - (s + u) \times n - (s + u)$ identity matrix, a notation that will be used subsequently. Furthermore, since we know that $f^u(p)$ is tangent to N'^u_p, we also have

$$\begin{aligned} D_z f_1(0; x) &= 0, \\ D_z f_2(0; x) &= 0. \end{aligned} \tag{5.74}$$

Based on (5.73)-(5.74), $DW|_{(x,0)}$ exists and equals

$$DW|_{(x,0)} = \begin{pmatrix} Id_{n-(s+u)} & 0 \\ Id_{n-(s+u)} & 0 \\ 0 & Id_u \\ 0 & 0 \end{pmatrix}. \tag{5.75}$$

Consequently, Σ is C^1 along its subset

$$M^* = \{(p,p) \,|\, p \in M\} \subset M \times \mathbb{R}^n, \tag{5.76}$$

which is just the diagonal embedding of M into $M \times \mathbb{R}^n$. At any point $p^* = (p,p) \in M^* \subset \Sigma$ the tangent space $T_{p^*}\Sigma$ is given by

$$T_{p^*}\Sigma = \text{Range}\,(DW|_{(x,0)}), \tag{5.77}$$

where x is the "tangential" coordinate of p on M. We easily see that M^* is an overflowing invariant manifold under the flow

$$\begin{aligned} \phi_t^* : M \times U &\to M \times \mathbb{R}^n, \\ \phi_t^*(p_1, p_2) &= (\phi_t(p_1), \phi_t(p_2)). \end{aligned}$$

We will show that Σ is a C^r submanifold of $M \times \mathbb{R}^n$ by arguing that Σ

is the unstable manifold of the C^r manifold M^* under the flow ϕ_t^*. To this end, we need to verify the hypotheses of the unstable manifold theorem for M^*, which requires the existence of stable and unstable subbundles N^{s*} and N^{u*}, respectively.

The diagonal embedding M^* of M into $M \times \mathbb{R}^n$ (given in (5.76)) does not tell us how to embed the bundles N^s and N^u into $T(M \times \mathbb{R}^n)$, hence we must do a separate construction here. It is enough to construct an embedding of $T\mathbb{R}^n|_M$ into $T(M \times \mathbb{R}^n)|_{M^*}$, because that restricts uniquely to embeddings of its subbundles. At any point $p \in M$, $T\mathbb{R}^n|_p$ is spanned by the tangent vectors of curves $\gamma_p : \mathbb{R} \to \mathbb{R}^n$, $s \mapsto \gamma_p(s)$, $\gamma_p(0) = p$. Therefore, it is enough to embed curves of this form into $M \times U$. But this can be done easily by defining $\gamma_{p^*}^* = e(\gamma_p)$ with

$$\gamma_{p^*}^*(s) = (p, \gamma_p(s)), \quad p^* = (p,p), \tag{5.78}$$

where e denotes the embedding map. Now $\gamma_{p^*}^*(s)$ has the property $\gamma_{p^*}^*(0) = p^*$. Using the tangent map of this embedding, we can then embed subbundles of $T\mathbb{R}^n|_M$ to obtain subbundles of $T(M \times \mathbb{R}^n)|_{M^*}$. In particular, we obtain the splitting

$$T(M \times \mathbb{R}^n)|_{M^*} = TM^* \oplus N^{u*} \oplus N^{s*},$$

which satisfies the hypotheses of the unstable manifold theorem. Indeed, one can easily check that the type numbers λ^{u*}, ν^{s*}, and σ^{s*} are the same when computed for this splitting as for the splitting $TM \oplus N^u \oplus N^s$. We conclude that M^* has a C^r unstable manifold, $W^u(M^*)$, which is tangent to $h_u^*(N^{u*})$ along M^*, with h_u^* defined analogously to h_u in Chapter 4.

The unstable manifold theorem guarantees that $W^u(M^*)$ has a unique Taylor expansion up to order r. Σ is clearly overflowing invariant by construction, so if we show that it is tangent to $h_u^*(N^{u*})$ along M^*, then it follows that Σ and $W^u(M^*)$ must have the same Taylor expansion up to order r; in particular, Σ is C^r. For this it is sufficient to show that $T_{p^*}\Sigma$ and $T_{p^*}M^* \oplus N_{p^*}^{u*}$ coincide for any $p^* \in M^*$.

A coordinate representation of the diagonal embedding of M is given by

$$
\begin{aligned}
\mathcal{M} : M &\to M \times \mathbb{R}^n, \\
x &\mapsto (x, x, 0_u, 0_s),
\end{aligned}
$$

therefore

$$
T_{p^*}M^* = \mathrm{Range}\,(D\mathcal{M})|_p = \mathrm{Range}\begin{pmatrix} Id_{n-(s+u)} \\ Id_{n-(s+u)} \\ 0 \\ 0 \end{pmatrix}. \tag{5.79}
$$

Selecting u curves, $\gamma_{p,1}(s), \ldots, \gamma_{p,u}(s)$, in $T_p\mathbb{R}^n$ which are tangent to the elements of the standard orthonormal basis of N'^u, respectively, (5.78) gives

$$
N_{p^*}^{u*} = \mathrm{Span}\,(\frac{d}{ds}\gamma_{p^*,i}^*(0))|_{i=1}^u = \mathrm{Range}\begin{pmatrix} 0 \\ 0 \\ Id_u \\ 0 \end{pmatrix}. \tag{5.80}
$$

Comparing (5.77), (5.79), and (5.80), we obtain that $T_{p^*}\Sigma = T_{p^*}M^* \oplus N_{p^*}^{u*}$, which concludes the proof.

PROOF OF PART 8

Part 8 is an immediate consequence of the preceeding results.

5.7   Persistence of the Fibers Under Perturbations

Exactly the same arguments can be applied as for $W^u(\bar{M})$ to show that the fibers persist and remain C^r under C^r perturbations.

6

Miscellaneous Properties and Results

In this chapter we collect together a number of useful results, as well as describe several ways the theory can be modified that are important for applications.

6.1   Inflowing Invariant Manifolds

Recall the definition of overflowing invariant manifolds given in Definition 3.0.1. From this definition it should be clear that under time reversal, i.e., $t \rightarrow -t$, an overflowing invariant manifold becomes an inflowing invariant manifold, and vice versa. Thus, if our vector field has an *inflowing* invariant manifold, then all of the previously developed theory can be applied to the *time-reversed* flow and its resulting *overflowing* invariant manifold. In this case, one only needs to take care in the characterization of stability of the inflowing invariant manifold. In particular, in applying the persistence theorem for overflowing invariant manifolds to inflowing invariant manifolds the conditions on the generalized Lyapunov-type numbers characterize an *unstable* inflowing invariant manifold, and the unstable manifold theorem for overflowing invariant manifolds becomes a *stable* manifold theorem for inflowing invariant manifolds.

6.2   Compact, Boundaryless Invariant Manifolds

Compact, boundaryless invariant manifolds are both overflowing and inflowing invariant, or according to the terminology of Definition 3.0.1, they are *invariant*. The persistence theorem, the unstable manifold theorem, the stable manifold theorem discussed above, and the stable and unstable foliation theorems can all be applied to compact, boundaryless invariant manifolds without modification.

6.3 Boundary Modifications

In certain cases, vector fields may possess invariant manifolds with boundary; however, the invariant manifolds are neither overflowing nor inflowing invariant. Rather, the vector field is either tangent to the the boundary or identically zero on the boundary. We give several examples to show that this is not such an uncommon occurrence, and then address the issue of how one deals with this situation in the context of the theory developed in the preceeding chapters.

A VECTOR FIELD NEAR A NONHYPERBOLIC FIXED POINT

Consider the following vector field

$$\dot{x} = -y + \epsilon f(x, y, z),$$

$$\dot{y} = x + \epsilon g(x, y, z), \tag{6.1}$$

$$\dot{z} = -z + \epsilon h(x, y, z), \qquad (x, y, z) \in \mathbb{R}^3.$$

At $\epsilon = 0$ the set

$$\bar{M} \equiv \left\{ (x, y, z) \in \mathbb{R}^3 \,|\, x^2 + y^2 \leq R, \, z = 0 \right\}$$

is a two-dimensional normally hyperbolic, attracting, invariant manifold with boundary. The vector field is tangent to the boundary since the boundary is an orbit.

A VECTOR FIELD WITH SLOWLY VARYING PARAMETERS

Consider the following *Duffing–van der Pol* oscillator:

$$\dot{x} = y,$$

$$\dot{y} = \alpha x - \beta x^2 y + \gamma x^3,$$

$$\dot{\alpha} = 0, \tag{6.2}$$

$$\dot{\beta} = 0,$$

$$\dot{\gamma} = 0, \qquad (x, y, \alpha, \beta, \gamma) \in \mathbb{R}^2 \times K,$$

where α, β, γ are parameters contained in some compact, connected set $K \subset \mathbb{R}^3$. Now suppose we consider a situation where the parameters have their own dynamics, i.e.,

$$\dot{x} = y,$$

$$\dot{y} = \alpha x - \beta x^2 y + \gamma x^3,$$

$$\dot{\alpha} = \epsilon f(x, y, \alpha, \beta, \gamma), \tag{6.3}$$

$$\dot{\beta} = \epsilon g(x, y, \alpha, \beta, \gamma),$$

$$\dot{\gamma} = \epsilon h(x, y, \alpha, \beta, \gamma), \qquad (x, y, \alpha, \beta, \gamma) \in \mathbb{R}^2 \times K.$$

At $\epsilon = 0$, (6.3) has an invariant manifold with boundary, denoted $\bar{M}$, given by

$$\bar{M} = \{(x, y, \alpha, \beta, \gamma) \, | \, x = y = 0\}.$$

For $\alpha > 0$ this invariant manifold is normally hyperbolic (it is actually

saddle type in stability). The boundary of $\bar{M}$ is given by the boundary of K and it is clear that at $\epsilon = 0$ the vector field is zero on the boundary.

HAMILTONIAN VECTOR FIELDS

Consider the following two-degree-of-freedom Hamiltonian system:

$$\dot{x} = \frac{\partial H}{\partial y}(x, y, I),$$

$$\dot{y} = -\frac{\partial H}{\partial x}(x, y, I),$$

$$\dot{I} = 0, \tag{6.4}$$

$$\dot{\theta} = \frac{\partial H}{\partial I}(x, y, I), \qquad (x, y, I, \theta) \in \mathbb{R} \times \mathbb{R} \times B \times S^1,$$

where B is some compact, connected set in $\mathbb{R}^+$. Let us assume that for all $I \in B$- the x-y component of (6.4) has a hyperbolic fixed point, denoted by $(\bar{x}(I), \bar{y}(I))$. Then the set

$$\bar{M} = \{(x, y, I, \theta) \, | \, x = \bar{x}(I), y = \bar{y}(I)), \ I \in B, \theta \in S^1\}$$

is a two-dimensional normally hyperbolic invariant manifold with boundary. The boundary is given by the two circles defined by $I \in \partial B$, $\theta \in S^1$, and the vector field is tangent to the boundary since each component of the the boundary is an orbit.

The only thing preventing the application of the theory developed in the previous chapters to these examples is that the vector field is tangent to the boundary, so that the vector field is neither overflowing nor inflowing invariant. However, this can be handled by considering a *modified* vector field. In particular, one can use "bump functions" to modify the vector field in an arbitrarily small neighborhood of the boundary so that it becomes either overflowing or inflowing invariant. The theory can then be applied to the modified vector field. Of course, generally the manifolds constructed in this way will depend on the nature of the modification at the boundary. However, trajectories that never pass through the modified region behave identically to those in the unmodified vector field. Orbits that pass through the modified region may have very different asymptotic behavior than they would have in the unmodified vector field. Specific examples where such "boundary modifications" are carried out can be found in Fenichel [1979] and Wiggins [1988].

Finally, it should be pointed out that the particular degeneracy of these examples (i.e., the fact that the vector field is zero or tangent on the boundary of the unperturbed normally hyperbolic invariant manifolds) makes it possible to prove the existence of both stable and unstable manifolds if the unperturbed manifold is of saddle stability type. This cannot be done if the unperturbed normally hyperbolic manifold is either overflowing or inflowing invariant and has a nonempty boundary.

6.4 Parameter-Dependent Vector Fields

Taking a cue from the above example, differentiabilty of invariant manifolds with respect to parameters can be deduced by extending the phase space by including the parameters as new dependent variables. We briefly outline the procedure.

Consider the vector field

$$\dot{x} = f(x, \mu), \qquad (x, \mu) \in \mathbb{R}^n \times K, \qquad (6.5)$$

where $K \subset \mathbb{R}^p$ is a compact, connected manifold with boundary. Suppose that (6.5) has an overflowing invariant manifold with boundary, $\bar{M}$. We consider the extended system

$$\dot{x} = f(x, \mu),$$

$$\dot{\mu} = 0, \qquad (x, \mu) \in \mathbb{R}^n \times K. \qquad (6.6)$$

Then it is readily verified that $\mathcal{M} \equiv \bar{M} \times K$ is an invariant manifold for (6.6) "with corners." In this case, one must "smooth out" the corners and modify the vector field on $\bar{M} \times \partial K$ as discussed in Section 6.3. Moreover, because the parameters do not change in time, the generalized Lyapunov-type numbers are unchanged for the enlarged system. Thus, the overflowing invariant manifold is C^r with respect to parameters.

6.5 Continuation of Overflowing Invariant Manifolds–The "Size" of the Perturbation

We now consider the issue of how large the perturbation may be with the overflowing invariant manifold still persisting. First, we will establish some notation. We have the unperturbed and perturbed C^r vector fields denoted by

$$\dot{x} = f(x), \qquad \dot{x} = f^{\text{pert}}(x), \qquad x \in \mathbb{R}^n,$$

having unperturbed and perturbed overflowing invariant manifolds denoted by

$$\bar{M}, \qquad \bar{M}^{\text{pert}}.$$

By construction, the perturbed manifold is the graph of a function

$$u^{\text{pert}} : M \to \mathbb{R}^n.$$

We denote the generalized Lyapunov-type numbers for the unperturbed overflowing invariant manifold by

$$\nu(p), \qquad \sigma(p),$$

and the generalized Lyapunov-type numbers for the perturbed overflowing invariant manifold by

$$\nu\left(u^{\text{pert}}(p)\right), \qquad \sigma\left(u^{\text{pert}}(p)\right).$$

We further define

$$\nu \equiv \sup_p \nu(p), \qquad \sigma \equiv \sup_p \sigma(p),$$

$$\nu^{\text{pert}} \equiv \sup_p \nu\left(u^{\text{pert}}(p)\right), \qquad \sigma^{\text{pert}} \equiv \sup_p \sigma\left(u^{\text{pert}}(p)\right).$$

It follows from the construction of the proofs of Theorem 3.3.1 that the overflowing invariant manifold will persist, with smoothness $C^{r'}$, for all vector fields C^1 close to the unperturbed vector field provided $\nu^{\text{pert}} < 1$ and and $\sigma^{\text{pert}} < \frac{1}{r'}$. Thus, the generalized Lyapunov-type numbers can be used as global bifurcation parameters.

6.6 Discrete Time Dynamics, or "Maps"

The theory developed in the previous chapters can be applied also to diffeomorphisms of $\mathbb{R}^n$. In fact, the basic results were developed in the context of the "time-T" map generated by the flow. One can compare the formulations of results in Fenichel [1971] with those in Fenichel [1974, 1977].

7

Examples

In this chapter we collect together several examples that illustrate the use and range of the theory developed in the previous chapters.

7.1 Invariant Manifolds Near a Hyperbolic Fixed Point

This example can be found in the thesis of Fenichel [1970]. It is the classical stable and unstable manifold theorem for a hyperbolic fixed point of a vector field. Consider the vector field

$$\dot{x} = Ax + f(x, y),$$

$$\dot{y} = By + g(x, y), \qquad (x, y) \in \mathbb{R}^u \times \mathbb{R}^s, \tag{7.1}$$

where $f(0,0) = g(0,0) = Df(0,0) = Dg(0,0) = 0$. A and B are constant matrices and we denote the eigenvalues of A and B by $\lambda_1, \lambda_2, \ldots, \lambda_u$ and $\mu_1, \mu_2, \ldots, \mu_s$, respectively. Furthermore, we assume

$$\text{Re } \lambda_1 \geq \text{Re } \lambda_2 \geq \cdots \geq \text{Re } \lambda_u > 0 > \text{Re } \mu_1 \geq \text{Re } \mu_2 \geq \cdots \geq \text{Re } \mu_s.$$

Hence, $(x, y) = (0, 0)$ is a hyperbolic fixed point.

Next, we consider the associated linear system

$$\dot{x} = Ax,$$

$$\dot{y} = By, \qquad (x, y) \in \mathbb{R}^u \times \mathbb{R}^s. \tag{7.2}$$

Under the above assumptions, for this equation the origin has an s-dimensional stable manifold (given by $x = 0$) and a u-dimensional unstable manifold (given by $y = 0$). Now for x and y small, (7.1) is a small C^1 perturbation of (7.2). We will show that (7.1) has an s-dimensional stable manifold tangent to $x = 0$ at the origin and a u-dimensional unstable manifold tangent to $y = 0$ at the origin. First, we deal with the unstable manifold.

We need a candidate for an overflowing invariant manifold for (7.2). If A is a diagonal matrix, the sphere of radius ϵ in the plane $y = 0$, i.e.,

$$\bar{M} \equiv \{(x, y) \mid \| x \| \leq \epsilon, \, y = 0\}$$

is easily shown to be overflowing invariant. If A is not diagonal, then one can use the theory of Jordan canonical forms to find an overflowing invariant ellipsoid; the details for this can be found in Arnold [1973]. We will similarly refer to the overflowing invariant manifold found through this procedure by $\bar{M}$.

The generalized Lyapunov-type numbers for the persistence theorem for overflowing invariant manifolds are easily computed and found to be

$$\nu(p) = e^{\mathrm{Re}\,\mu_1}, \qquad \sigma(p) = \frac{\mathrm{Re}\,\lambda_u}{\mathrm{Re}\,\mu_1}, \qquad \forall p \in \bar{M}.$$

(Note: The generalized Lyapunov-type numbers are constant since the "unperturbed" vector field is linear.) Hence, the hypotheses of Theorem 3.3.1 hold. Thus, for ϵ sufficiently small, (7.1) has an overflowing invariant manifold tangent to $y = 0$ at the origin. Moreover, since $\sigma(p)$ is negative, the overflowing invariant manifold is as differentiable as the vector field. The same argument can be applied to the time-reversed vector field to obtain the existence of the stable manifold.

7.2   Invariant Manifolds Near a Nonhyperbolic Fixed Point

Consider the vector field

$$\dot{x} = Ax + f(x, y),$$

$$\dot{y} = By + g(x, y), \qquad (x, y) \in \mathbb{R}^c \times \mathbb{R}^s, \tag{7.3}$$

where $f(0,0) = g(0,0) = Df(0,0) = Dg(0,0) = 0$. A and B are constant

matrices and we denote the eigenvalues of A and B by $\lambda_1, \lambda_2, \ldots, \lambda_u$ and $\mu_1, \mu_2, \ldots, \mu_s$, respectively. Furthermore, we assume

$$0 > \mathrm{Re}\,\mu_1 \geq \mathrm{Re}\,\mu_2 \geq \cdots \geq \mathrm{Re}\,\mu_s, \qquad \text{and} \quad \mathrm{Re}\,\lambda_i = 0,\, i = 1, \ldots, c.$$

Hence, $(x, y) = (0, 0)$ is a nonhyperbolic fixed point.

Next, we consider the associated linear system

$$\dot{x} = Ax,$$

$$\dot{y} = By. \qquad (x, y) \in \mathbb{R}^c \times \mathbb{R}^s. \tag{7.4}$$

Under the above assumptions, for this equation the origin has an s-dimensional stable manifold (given by $x = 0$) and a c-dimensional center manifold (given by $y = 0$). As above, for x and y small, (7.3) is a small C^1

perturbation of (7.4). Our goal is to show that the center manifold exists for the nonlinear vector field by using the persistence theorem for overflowing invariant manifolds. However, because the eigenvalues of A have zero real parts, $\bar{M}$ as defined above is not necessarily an overflowing invariant manifold. This problem can be overcome by modifying the vector field in a small neighborhood of the boundary of M so that it becomes overflowing invariant.

Let us take a C^∞ "bump" function, χ, defined as

$$\chi : [0, \infty) \rightarrow \mathbb{R},$$

having the following properties

1. $0 \le \chi(v) \le \delta, \quad \forall v \ge 0,$

2. $\chi(v) = 0, \quad \forall v \in [0, \epsilon - \eta],$

3. $\chi(v) = \delta, \quad \forall v \in [\epsilon, \epsilon + \frac{\eta}{2}],$

4. $\chi(v) = 0, \quad \forall v \in [\epsilon + \eta, \infty),$

where η can be taken arbitrarily small.

We now modify (7.4) as follows:

$$\dot{x} = Ax + \chi(\| x \|)x,$$

$$\dot{y} = By, \qquad (x, y) \in \mathbb{R}^c \times \mathbb{R}^s,$$

(7.5)

and choose δ so that

$$\frac{\langle x, Ax \rangle}{\| x \|^2} + \delta > 0 \qquad \text{for } \| x \| = \epsilon.$$

In this case

$$\bar{M} \equiv \{(x, y) \mid \| x \| \le \epsilon, \, y = 0\}$$

becomes an overflowing invariant manifold for (7.5). Moreover, (7.4) and (7.5) are identical, except in an arbitrarily small neighborhood of the boundary of M. Hence, the arguments of the previous example go through in exactly the same way for the modified vector field in this example. We note that the persisting overflowing invariant manifold (i.e., the "center manifold" of the origin) *depends* on the nature of the modification at the boundary of M. Moreover, trajectories which pass through the modified region may have very different asymptotic behaviors compared to those trajectories in the unmodified vector field.

7.3   Weak Hyperbolicity

Consider the C^r vector field

$$\dot{x} = \epsilon f(x) + \epsilon^2 g(x, \epsilon), \qquad x \in \mathbb{R}^n,$$

where ϵ is viewed as a small perturbation parameter. Let us suppose that the vector field $\dot{x} = \epsilon f(x)$ has an overflowing invariant manifold, $\bar{M}$, with generalized Lyapunov-type numbers that satisfy the hypotheses of Theorem 3.3.1. We would like to apply Theorem 3.3.1 and argue that $\epsilon^2 g(x)$ is a C^1 perturbation of $\epsilon f(x)$ and that $\bar{M}$ persists for the perturbed vector field. However, there is a difficulty with this argument which must be faced first, namely, the generalized Lyapunov-type number $\nu(p)$ is $\mathcal{O}(\epsilon)$. Thus, as the magnitude of the perturbation goes to zero, the hyperbolicity also goes to zero. This is what we mean by the term "weak hyperbolicity." The problem can be stated in another way. If one fixes the strength of the hyperbolicity (as measured by the generalized Lyapunov-type numbers), then the size of the perturbation is also fixed.

This situation arises frequently in normal form type analyses where one has a normally hyperbolic invariant manifold in the truncated normal form and one wants to argue that it persists when the influence of the "tail" of the normal form is included.

This problem is easy to deal with in this particular example since ϵ multiplies the entire vector field. In this case we rescale time by letting $t \rightarrow \frac{t}{\epsilon}$ so that the vector field then takes the form

$$\dot{x} = f(x) + \epsilon g(x), \qquad x \in \mathbb{R}^n.$$

The theory developed in the previous chapters can now be applied immediately to the vector field in this form, and we obtain that $\bar{M}$ persists for $\epsilon > 0$ sufficiently small.

Difficulties arise when the perturbation parameter does not multiply the entire vector field, but only certain components. Let us consider a common case that illustrates this issue.

Consider the vector field

$$\dot{x} = \epsilon f(x) + \epsilon^2 g(x, \theta, \epsilon),$$

$$\dot{\theta} = \Omega(x) + \epsilon h(x) + \epsilon^2 k(x, \theta, \epsilon), \qquad (x, \theta) \in \mathbb{R}^n \times T^m. \tag{7.6}$$

Suppose that at $x = x_0$ $f(x_0) = 0$ and all eigenvalues of $Df(x_0)$ lie in the left half-plane. Then the vector field

$$\dot{x} = \epsilon f(x),$$

$$\dot{\theta} = \Omega(x) + \epsilon h(x), \qquad (x, \theta) \in \mathbb{R}^n \times T^m, \tag{7.7}$$

has a normally hyperbolic, attracting, invariant torus, T, given by

$$T = \{(x, \theta) \mid x = x_0\},$$

and the trajectories on the torus are given by

$$
\begin{aligned}
x(t) &= x_0, && (7.8) \\
\theta(t) &= (\Omega(x_0) + \epsilon h(x_0))\, t + \theta_0. && (7.9)
\end{aligned}
$$

We are in a setting where the persistence theorem for overflowing invariant manifolds can be applied. (Note: The fact that the torus is boundaryless is irrelevant for this application.) We want to argue that for ϵ sufficiently small, the attracting invariant torus in (7.7) persists in (7.6). To do this, we use an idea due to Kopell [1985].

Consider the auxiliary system

$$
\begin{aligned}
\dot{x} &= \delta f(x) + \epsilon^2 g(x, \theta, \epsilon), \\
&&&&& (7.10) \\
\dot{\theta} &= \Omega(x) + \delta h(x) + \epsilon^2 k(x, \theta, \epsilon), && (x, \theta) \in \mathbb{R}^n \times T^m,
\end{aligned}
$$

where δ is regarded as fixed. We know from the persistence theorem for overflowing invariant manifolds (Theorem 3.3.1) that for a fixed δ, there exists ϵ sufficiently small such that (7.10) has a normally hyperbolic invariant torus, T^ϵ. We want to argue that for δ sufficiently small, ϵ can be increased to the size of δ, i.e., we can take $\epsilon \leq \delta$ and still have the persisting torus T^ϵ.

First, we recall some notation and, in that context, establish some new notation. By construction, the perturbed manifold is the graph of a function

$$u^\epsilon : T^\delta \to \mathbb{R}^n,$$

where T^δ denotes the unperturbed torus, and we also have

$$\| A_t(p) \| \equiv \| D\phi_{-t}|_{T^\delta}(p) \|,$$

$$\| A_t^\epsilon(u^\epsilon(p)) \| \equiv \| D\phi_{-t}^\epsilon|_{T^\epsilon}(u^\epsilon(p)) \|,$$

$$\qquad (7.11)$$

$$\| B_t(p) \| \equiv \| \Pi D\phi_t(\phi_{-t}(p)) \|,$$

$$\| B_t^\epsilon(u^\epsilon(p)) \| \equiv \| \Pi^\epsilon D\phi_t^\epsilon(\phi_{-t}^\epsilon(u^\epsilon(p))) \|.$$

The generalized Lyapunov-type numbers of Chapter 3 for the unperturbed overflowing invariant manifold are given by

$$\nu(p) = \overline{\lim}_{t \to \infty} \| B_t(p) \|^{\frac{1}{t}} = e^{\delta\, \mathrm{Re}\, \lambda_{\min}},$$

$$\sigma(p) = \overline{\lim}_{t \to \infty} \frac{\| A_t(p) \|}{- \| B_t(p) \|} = 0,$$

where $\lambda_{\min}$ denotes the eigenvalue of $Df(x_0)$ with the largest real part. The generalized Lyapunov-type numbers for the perturbed overflowing invariant manifold are given by

$$\nu\left(u^\epsilon(p)\right) = \overline{\lim}_{t\to\infty} \parallel B_t^\epsilon\left(u^\epsilon(p)\right) \parallel^{\frac{1}{t}},$$

$$\sigma\left(u^\epsilon(p)\right) = \overline{\lim}_{t\to\infty} \frac{\parallel A_t^\epsilon\left(u^\epsilon(p)\right) \parallel}{-\parallel B_t^\epsilon\left(u^\epsilon(p)\right) \parallel}.$$

We now want to obtain an upper bound on $\nu\left(u^\epsilon(p)\right)$. We write $t = nT + r$, where $n \in \mathbb{N}$ and $r \in [0, T)$. We then have the inequalities

$$
\begin{aligned}
\parallel B_t^\epsilon(p) \parallel^{\frac{1}{t}} &= \parallel B_{nT+r}^\epsilon(p) \parallel^{1/(nT+r)} \\
&\leq \parallel B_{nT}^\epsilon(p) \parallel^{1/(nT+r)} \parallel B_r^\epsilon(p) \parallel^{1/t} \\
&\leq \parallel B_T^\epsilon(p) \parallel^{1/(T+r/n)} \parallel B_r^\epsilon(p) \parallel^{1/t}.
\end{aligned}
$$

$$(7.12)$$

Taking the limsup of this expression as $t \to \infty$ gives

$$\nu\left(u^\epsilon(p)\right) \leq \parallel B_T^\epsilon(p) \parallel^{\frac{1}{T}}.$$

The generalized Lyapunov-type numbers are not typically differentiable (or even continuous) functions of parameters. Therefore, the significance of this inequality is that although the left-hand side of the inequality may not be continuous in ϵ, the right-hand side is C^{r-1} in ϵ. Thus, we have

$$\nu\left(u^\epsilon(p)\right) \leq \parallel B_T(p) \parallel^{\frac{1}{T}} + \mathcal{O}(\epsilon) = e^{\delta \operatorname{Re} \lambda_{\min}} + \mathcal{O}(\epsilon).$$

Thus, if δ is originally chosen small enough, then we have

$$\nu\left(u^\epsilon(p)\right) \leq e^{\delta \operatorname{Re} \lambda_{\min}} + \mathcal{O}(\epsilon) \leq e^{\delta \operatorname{Re} \lambda_{\min}} + \mathcal{O}(\delta) < 1$$

for all $\epsilon \leq \delta$. A similar argument reveals that $\sigma\left(u^\epsilon(p)\right) < 1$ holds for all $\epsilon \leq \delta$. Thus, as discussed in Section 6.5, the invariant manifold T^ϵ continues to exist for $0 < \epsilon \leq \delta$ for system (7.6).

7.4   Asymptotic Expansions for Invariant Manifolds

Once the existence and smoothness of an invariant manifold has been established, Taylor expansion methods can be used for its approximation. Specific examples can be found in Fenichel [1979], as well as in the example in the next section.

7.5 The Invariant Manifold Structure Associated with the Study of Orbits Homoclinic to Resonances

In this section we will look at some aspects of the invariant manifold structure associated with a class of perturbed two-degree-of-freedom Hamiltonian systems having the following form:

Non-Hamiltonian Perturbations

$$
\begin{aligned}
\dot{x} &= JD_xH(x,I;\mu) + \epsilon g^x(x,I,\theta;\mu,\epsilon), \\
\dot{I} &= \epsilon g^I(x,I,\theta;\mu,\epsilon), \qquad\quad (x,I,\theta) \in \mathbb{R}^2 \times \mathbb{R} \times S^1, \\
\dot{\theta} &= D_IH(x,I;\mu) + \epsilon g^\theta(x,I,\theta;\mu,\epsilon). \qquad\qquad (7.13)
\end{aligned}
$$

Hamiltonian Perturbations

$$
\begin{aligned}
\dot{x} &= JD_xH(x,I;\mu) + \epsilon JD_xH_1(x,I,\theta;\mu,\epsilon), \\
\dot{I} &= -\epsilon D_\theta H_1(x,I,\theta;\mu,\epsilon), \qquad\quad (x,I,\theta) \in \mathbb{R}^2 \times \mathbb{R} \times S^1, \\
\dot{\theta} &= D_IH(x,I;\mu) + \epsilon D_IH_1(x,I,\theta;\mu,\epsilon), \qquad\qquad (7.14)
\end{aligned}
$$

where J denotes the usual symplectic matrix, i.e.,

$$
\begin{pmatrix} 0 & 1 \\ -1 & 0 \end{pmatrix};
$$

all functions are sufficiently differentiable ($C^r, r \geq 3$ is suffficient) on the domains of interest, $0 \leq \epsilon << 1$ is the perturbation parameter, and $\mu \in V \subset \mathbb{R}^p$ is a vector of parameters. (Note: D_x, etc., will denote partial derivatives and d/dx, etc., will denote total derivatives.)

In Kovačič and Wiggins [1992], Haller and Wiggins [1993a], [1993b], and McLaughlin et al. [1993], global perturbation methods are developed for the study of homoclinic and heteroclinic orbits that connect different types of invariant sets in the phase space of (7.13) and (7.14). A central feature of these methods is the use of the geometrical structure of the integrable Hamiltonian unperturbed problem in order to develop appropriate "coordinates" for studying the perturbed problem. The purpose of this section is not to develop these global perturbation methods, but rather to explore more fully the invariant manifold structure of these systems. We begin by stating our assumptions on the unperturbed problem.

7.5.1 THE ANALYTIC AND GEOMETRIC STRUCTURE OF THE UNPERTURBED EQUATIONS

The unperturbed equations are given by

$$
\begin{aligned}
\dot{x} &= JD_xH(x,I;\mu), \\
\dot{I} &= 0, \qquad (x,I,\theta;\mu) \in \mathbb{R}^2 \times \mathbb{R} \times S^1 \times V, \\
\dot{\theta} &= D_IH(x,I;\mu), \tag{7.15}
\end{aligned}
$$

Note the simple structure of (7.15); effectively, it is two uncoupled one-degree-of-freedom Hamiltonian systems. Since $\dot{I} = 0$, the I variable enters the x component of (7.15) only as a parameter. Hence, the x component of (7.15) can be solved independently since it is just a one-parameter family of one-degree-of-freedom (hence integrable) Hamiltonian systems (suppressing the external μ parameter dependence). This solution can then be substituted into the θ component of (7.15) which can then be integrated to yield the full solution. We make the following assumption on the x component of (7.15).

Assumption 1. For all $I \in [I_1, I_2]$, $\mu \in V$, the equation

$$
\dot{x} = JD_xH(x,I;\mu) \tag{7.16}
$$

has a hyperbolic fixed point, $\tilde{x}_0(I;\mu)$, connected to itself by a homoclinic trajectory, $x^h(t,I;\mu)$, i.e., $\lim_{t \to \pm\infty} x^h(t,I;\mu) = \tilde{x}_0(I;\mu)$.

We now want to use the simple structure of the "decoupled" unperturbed system to build up a picture of the geometry in the full four-dimensional phase space. This will provide the framework for studying the perturbed problem which will be fully four dimensional.

Assumption 1 implies that in the full four-dimensional phase space, the set

$$
\mathcal{M} = \{(x,I,\theta) \mid x = \tilde{x}_0(I;\mu),\ I_1 \leq I \leq I_2,\ 0 \leq \theta < 2\pi,\ \mu \in V\} \tag{7.17}
$$

is a two-dimensional, invariant manifold. Moreover, the hyperbolic saddle-type fixed point from Assumption 1 gives rise to *normal hyperbolicity* of $\mathcal{M}$.

The two-dimensional, normally hyperbolic, invariant manifold $\mathcal{M}$ has three-dimensional stable and unstable manifolds which we denote as $W^s(\mathcal{M})$ and $W^u(\mathcal{M})$, respectively. This can be inferred from the structure of the x component of (7.15) given in Assumption 1. Moreover, the existence of the homoclinic orbit of the x component of (7.15) implies that $W^s(\mathcal{M})$ and $W^u(\mathcal{M})$ intersect (nontransversely) along a three-dimensional *homoclinic manifold* which we denote by Γ. A trajectory in

$\Gamma \equiv W^s(\mathcal{M}) \cap W^u(\mathcal{M})$ can be expressed as

$$\left(x^h(t, I; \mu), \ I, \ \theta(t, I, \theta_0; \mu) = \int_0^t D_I H(x^h(s, I; \mu), I; \mu) ds + \theta_0 \right), \quad (7.18)$$

and it is clear that this trajectory approaches $\mathcal{M}$ as $t \to \pm\infty$ since $x^h(t, I; \mu) \to \tilde{x}_0(I; \mu)$ as $t \to \pm\infty$.

THE DYNAMICS OF THE UNPERTURBED SYSTEM RESTRICTED TO $\mathcal{M}$

The unperturbed system restricted to $\mathcal{M}$ is given by

$$\begin{aligned} \dot{I} &= 0, \\ \dot{\theta} &= D_I H(\tilde{x}_0(I; \mu), I; \mu), \qquad I_1 \leq I \leq I_2. \end{aligned} \qquad (7.19)$$

Thus, if $D_I H(\tilde{x}_0(I; \mu), I; \mu) \neq 0$, then $I = $ constant labels a periodic orbit, and if $D_I H(\tilde{x}_0(I; \mu), I; \mu) = 0$, then $I = $ constant labels a circle of fixed points. We refer to a value of I for which $D_I H(\tilde{x}_0(I; \mu), I; \mu) = 0$ as a *resonant I value* and these fixed points as *resonant fixed points*. We make the following assumption on the unperturbed system restricted to $\mathcal{M}$.

Assumption 2 : Resonance. There exists a value of $I \in [I_1, I_2]$, denoted I^r, at which $D_I H(\tilde{x}_0(I^r; \mu), I^r; \mu) = 0$.

THE DYNAMICS IN Γ AND ITS RELATION TO THE DYNAMICS IN $\mathcal{M}$

Recall the expression for an orbit in Γ given in (7.18). As $x^h(t, I; \mu) \to \tilde{x}_0(I; \mu)$ and I remains constant, we want to call attention to the expression that we will define as

$$\triangle\theta(I, \mu) = \theta(+\infty, I, \theta_0; \mu) - \theta(-\infty, I, \theta_0; \mu)$$

$$= \int_{-\infty}^{+\infty} D_I H\left(x^h(t, I; \mu), I; \mu\right) dt. \qquad (7.20)$$

Now for an I value such that $D_I H(\tilde{x}_0(I; \mu), I; \mu) \neq 0$ it is easy to see that $\triangle\theta(I, \mu)$ is not finite. This just reflects the fact that asymptotically the orbit approaches a periodic orbit whose phase constantly changes forever. However, at resonant I values, $\triangle\theta$ is finite since the integral in (7.20) converges. (The convergence of the integral follows from the fact that $x^h(t, I; \mu) \to \tilde{x}_0(I; \mu)$ exponentially fast as $t \to \pm\infty$; hence, at resonance $D_I H(x^h(t, I; \mu), I; \mu)$ goes to zero exponentially fast as t goes to $\pm\infty$.) Since resonant I values determine circles of fixed points on $\mathcal{M}$, the orbit $(x^h(t, I^r; \mu), I^r, \theta(t, I^r, \theta_0; \mu))$ is typically a heteroclinic connection between different points on the resonant circle of fixed points. (The connection will be homoclinic if $\triangle\theta(I^r, \mu) = 2\pi n$, for some integer n.) The number $\triangle\theta(I^r, \mu)$ gives the shift in phase between the two endpoints of the heteroclinic trajectory along the circle of fixed points.

In Figs. 7.1 and 7.2 we illustrate the relevant aspects of the geometry and dynamics of the unperturbed system.

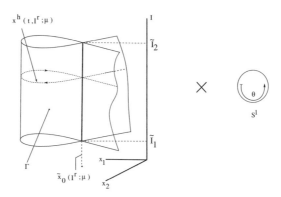

FIGURE 7.1. The Invariant Manifold Structure of the Unperturbed System.

7.5.2   THE ANALYTIC AND GEOMETRIC STRUCTURE OF THE PERTURBED EQUATIONS

The geometrical structure of the unperturbed system will provide us with the framework for understanding certain types of global behavior that can occur in the perturbed system. In particular, $\mathcal{M}$ along with its stable and unstable manifolds will persist in the perturbed system; however, the dynamics on these manifolds will be quite different. This should be evident since $\mathcal{M}$ contains a circle of fixed points and the stable and unstable manifolds of $\mathcal{M}$ have a coinciding branch; both of these structures are highly degenerate.

THE PERSISTENCE OF $\mathcal{M}$ AND ITS STABLE AND UNSTABLE MANIFOLDS

As we have discussed in Sections 6.3 and 7.2, the problem of the persistence of invariant manifolds with boundary under perturbations gives rise to certain technical questions concerning the nature of the trajectories at the boundary. In order to address these questions precisely for this problem, we begin by defining the set

$$U^\delta = \{(x, I, \theta) \mid |x - \tilde{x}_0(I; \mu)| \le \delta, \ \tilde{I}_1 \le I \le \tilde{I}_2\}, \tag{7.21}$$

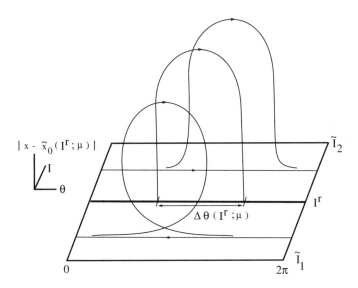

FIGURE 7.2. An Illustration of the Orbits Homoclinic to the Different Invariant Sets in $\mathcal{M}$.

where

$$I_1 \leq \tilde{I}_1 < \tilde{I}_2 \leq I_2.$$

If $I_1 = \tilde{I}_1$ and $I_2 = \tilde{I}_2$, then clearly U^δ is the closure of a neighborhood of $\mathcal{M}$. However, in order to modify the vector field at the boundary to make it overflowing or inflowing invariant, we will need to slightly restrict the range of I values in discussing the perturbed manifolds and it is for this reason that the I interval in the definition of U^δ has been restricted. (Note: $\tilde{I}_1$ can be chosen arbitrarily close to I_1 and $\tilde{I}_2$ can be chosen arbitrarily close to I_2.) The set U^δ will be useful in characterizing the nature of trajectories near the invariant manifolds.

For the unperturbed system, we define the local stable and unstable manifolds of $\mathcal{M}$ as follows:

$$W_{loc}^s(\mathcal{M}) \equiv W^s(\mathcal{M}) \cap U^\delta, \tag{7.22}$$

$$W_{loc}^u(\mathcal{M}) \equiv W^u(\mathcal{M}) \cap U^\delta. \tag{7.23}$$

We now state a persistence theorem which is a restatement of Theorems 3.3.1 and 4.5.1, combined with our discussions in Sections 6.1, 6.3, and 6.4.

Theorem 7.5.1 *There exists $\epsilon_0 > 0$ sufficiently small such that for $0 < \epsilon \leq \epsilon_0$, $\mathcal{M}$ persists as a C^r, locally invariant, two-dimensional, normally hyperbolic manifold with boundary, which we denote by $\mathcal{M}_\epsilon$, having the following properties:*

1. *$\mathcal{M}_\epsilon$ is C^r in ϵ and μ.*

2. *$\mathcal{M}_\epsilon$ is C^r ϵ-close to $\mathcal{M}$ and can be represented as a graph over $\mathcal{M}$ as*

$$\mathcal{M}_\epsilon = \{(x, I, \theta) \mid x = \tilde{x}_\epsilon(I, \theta; \mu) = \tilde{x}_0(I; \mu) + \epsilon\tilde{x}_1(I, \theta; \mu) + \mathcal{O}(\epsilon^2),$$

$$\tilde{I}_1 \leq I \leq \tilde{I}_2, \quad \theta \in S^1\}.$$

(7.24)

Moreover, there exists δ_0 sufficiently small (depending on ϵ) such that for $0 < \delta < \delta_0$ there exists locally invariant manifolds in U^δ, denoted $W^s_{loc}(\mathcal{M}_\epsilon)$, $W^u_{loc}(\mathcal{M}_\epsilon)$, having the following properties:

3. *$W^s_{loc}(\mathcal{M}_\epsilon)$ and $W^u_{loc}(\mathcal{M}_\epsilon)$ are C^r in ϵ and μ.*

4. *$W^s_{loc}(\mathcal{M}_\epsilon) \cap W^u_{loc}(\mathcal{M}_\epsilon) = \mathcal{M}_\epsilon$.*

5. *$W^s_{loc}(\mathcal{M}_\epsilon)$ (resp. $W^u_{loc}(\mathcal{M}_\epsilon)$) is a graph over $W^s_{loc}(\mathcal{M})$ (resp. $W^u_{loc}(\mathcal{M})$) and is C^r ϵ-close to $W^s_{loc}(\mathcal{M})$ (resp. $W^u_{loc}(\mathcal{M})$).*

6. *Let $y^s_\epsilon(t) \equiv (x^s_\epsilon(t), I^s_\epsilon(t), \theta^s_\epsilon(t))$ (resp. $y^u_\epsilon(t) \equiv (x^u_\epsilon(t), I^u_\epsilon(t), \theta^u_\epsilon(t))$) denote a trajectory that is in $W^s_{loc}(\mathcal{M}_\epsilon)$ (resp. $W^u_{loc}(\mathcal{M}_\epsilon)$) at $t = 0$. Then as $t \to +\infty$ (resp. $t \to -\infty$) either*

   (a) *$y^s_\epsilon(t) \equiv (x^s_\epsilon(t), I^s_\epsilon(t), \theta^s_\epsilon(t))$ (resp. $y^u_\epsilon(t) \equiv (x^u_\epsilon(t), I^u_\epsilon(t), \theta^u_\epsilon(t))$) crosses ∂U^δ*

   or

   (b) *$\lim_{t \to \infty} d\left(y^s_\epsilon(t), \mathcal{M}_\epsilon\right) = 0$ (resp. $\lim_{t \to -\infty} d\left(y^u_\epsilon(t), \mathcal{M}_\epsilon\right) = 0$), where $d(\cdot, \cdot)$ denotes the standard metric in the phase space.*

We refer to $W^s_{loc}(\mathcal{M}_\epsilon)$ and $W^u_{loc}(\mathcal{M}_\epsilon)$ as the local stable and unstable manifolds of $\mathcal{M}_\epsilon$, respectively.

We make the following remarks concerning the consequences and implications of this theorem.

Remark 1. The term *locally invariant* means that trajectories with initial conditions on $\mathcal{M}_\epsilon$ may leave $\mathcal{M}_\epsilon$; however, they may do so only by crossing the boundary of $\mathcal{M}_\epsilon$. In proving the persistence of $\mathcal{M}$ under perturbation it is necessary to know the stability properties of trajectories in $\mathcal{M}$ on semi-infinite time intervals. Technically, this control is accomplished by modifying the unperturbed vector field (7.13) in an arbitrarily small neighborhood of the boundary of $\mathcal{M}$ by using C^∞ "bump functions"; this procedure is explained in Wiggins [1988] and in Section 6.3. The perturbed manifold is then constructed as a graph over the unperturbed manifold by using the graph transform technique. This is the reason why the range of I values for which $\mathcal{M}_\epsilon$ exists in the perturbed vector field (7.13) may need to be slightly decreased.

Remark 2. We define the global stable and unstable manifolds of $\mathcal{M}_\epsilon$, denoted $W^s(\mathcal{M}_\epsilon)$ and $W^u(\mathcal{M}_\epsilon)$, respectively, as follows. Let $\phi_t(\cdot)$ denote the flow generated by (7.13); then we define

$$W^s(\mathcal{M}_\epsilon) = \bigcup_{t \leq 0} \phi_t \left(W^s_{loc}(\mathcal{M}_\epsilon) \cap U^\delta \right),$$

$$W^u(\mathcal{M}_\epsilon) = \bigcup_{t \geq 0} \phi_t \left(W^u_{loc}(\mathcal{M}_\epsilon) \cap U^\delta \right).$$

(7.25)

Remark 3. The phrase *stable manifold of an invariant set* typically means the manifold of trajectories that approach the invariant set as $t \to +\infty$. However, our definition has a slightly different meaning that is peculiar to our invariant set, i.e., $\mathcal{M}_\epsilon$, having a boundary. This is characterized in terms of the alternatives (a) and (b) of part 6 of Theorem 7.5.1; similarly for the *unstable manifold of an invariant set*.

APPROXIMATE CALCULATION OF $\mathcal{M}_\epsilon$ NEAR RESONANCE

The fact that $\mathcal{M}_\epsilon$ is C^r ($r \geq 2$) in μ and ϵ allows us to Taylor expand the manifold in powers of μ and ϵ. This will be important because we may need to explicitly compute the $\mathcal{O}(\epsilon)$ term in the expansion of $\tilde{x}_\epsilon(I, \theta; \mu)$ given in (7.24). We now explain how this may be done.

The following calculations are independent of whether or not the perturbation is Hamiltonian. Therefore, we give the calculations using the notation of a general perturbation.

Differentation of $\tilde{x}_\epsilon(I, \theta; \mu)$ along the perturbed vector field (7.13) gives a quasilinear partial differential equation that $\tilde{x}_\epsilon(I, \theta; \mu)$ must satisfy. This equation is given by

$$JD_x H(\tilde{x}_\epsilon, I; \mu) + \epsilon g^x(\tilde{x}_\epsilon, I, \theta; \mu, \epsilon)$$

$$= \epsilon \left(D_I \tilde{x}_\epsilon \right) g^I(\tilde{x}_\epsilon, I, \theta; \mu, \epsilon) + (D_\theta \tilde{x}_\epsilon) \left(D_I H(\tilde{x}_\epsilon, I; \mu) + \epsilon g^\theta(\tilde{x}_\epsilon, I, \theta; \mu, \epsilon) \right).$$

(7.26)

We can differentiate (7.26) with respect to ϵ and obtain equations that the derivatives of $\tilde{x}_\epsilon(I, \theta; \mu)$ must satisfy. In this way, equating $\mathcal{O}(\epsilon)$ terms on both sides, we find that $\tilde{x}_1(I, \theta; \mu)$ must satisfy the following ordinary differential equation:

$$
\begin{aligned}
&-\left(D_\theta \tilde{x}_1\right) D_I H(\tilde{x}_0(I; \mu), I; \mu) + J D_x^2 H(\tilde{x}_0(I; \mu), I; \mu) \tilde{x}_1 \\
&= \left(D_I \tilde{x}_0(I; \mu)\right) g^I(\tilde{x}_0(I; \mu), I, \theta; \mu, 0) - g^x(\tilde{x}_0(I; \mu), I, \theta; \mu, 0).
\end{aligned}
\tag{7.27}
$$

We want to point out that in one special circumstance (indeed, the situation that will be most important to us) the solution of (7.27) can immediately be written down, namely, at resonance. For at resonance, i.e., $I = I^r$, we have $D_I H(\tilde{x}_0(I^r; \mu), I^r; \mu) = 0$ so that (7.27) reduces to an algebraic equation with solution

$$
\begin{aligned}
\tilde{x}_1 \;=\; &\left(J D_x^2 H(\tilde{x}_0(I^r; \mu), I^r)\right)^{-1} \left(D_I \tilde{x}_0(I^r; \mu) g^I(\tilde{x}_0(I^r; \mu), I^r, \theta; \mu, 0)\right. \\
&\left. - g^x(\tilde{x}_0(I^r; \mu), I^r, \theta; \mu, 0)\right).
\end{aligned}
\tag{7.28}
$$

It is also easy to find an expression for $D_I \tilde{x}_0(I; \mu)$ by implicitly differentiating the equation $D_x H(\tilde{x}_0(I; \mu), I; \mu) = 0$. This simple calculation gives

$$
D_I \tilde{x}_0(I; \mu) = -\left(D_x^2 H(\tilde{x}_0(I; \mu), I; \mu)\right)^{-1} \left(D_I D_x H(\tilde{x}_0(I; \mu), I; \mu)\right). \tag{7.29}
$$

Equations (7.28) and (7.29) will be useful later. We remark that the invertibilty of $D_x^2 H(\tilde{x}_0(I; \mu), I; \mu)^{-1}$ follows from the hyperbolicity of $\tilde{x}_0(I; \mu)$.

The Fibering of $W^s(\mathcal{M}_\epsilon)$ and $W^u(\mathcal{M}_\epsilon)$ Near Resonance: The Singular Perturbation Nature

In this subsection we want to examine the foliation of $W^s(\mathcal{M}_\epsilon)$ and $W^u(\mathcal{M}_\epsilon)$ as described in Chapter 5 and Section 6.1. We will study these foliations in a neighborhood of the hypersurface $I = I^r$.

More insight into these questions can be obtained by directly examining the equations of motion. *For this discussion we will restrict ourselves to non-Hamiltonian perturbations; the details for Hamiltonian perturbations can be found in Haller and Wiggins [1993a].*

First, we must do some preliminary transformations of the equations. The perturbed vector field restricted to $\mathcal{M}_\epsilon$ is given by

$$
\begin{aligned}
\dot{I} &= \epsilon g^I(\tilde{x}_\epsilon(I, \theta; \mu), I, \theta; \mu, \epsilon), \\
\dot{\theta} &= D_I H(\tilde{x}_\epsilon(I, \theta; \mu), I; \mu) + \epsilon g^\theta(\tilde{x}_\epsilon(I, \theta; \mu), I, \theta; \mu, \epsilon).
\end{aligned}
\tag{7.30}
$$

Taylor expanding (7.30) in powers of ϵ gives

$$
\begin{aligned}
\dot{I} &= \epsilon g^I(\tilde{x}_0(I;\mu), I, \theta; \mu, 0) + \epsilon^2 \left(\langle D_x g^I(\tilde{x}_0(I;\mu), I, \theta; \mu, 0), \tilde{x}_1(I, \theta; \mu) \rangle \right. \\[2mm]
&\quad \left. + D_\epsilon g^I(\tilde{x}_0(I;\mu), I, \theta; \mu, 0) \right) + \mathcal{O}(\epsilon^3), \\[3mm]
\dot{\theta} &= D_I H(\tilde{x}_0(I;\mu), I; \mu) + \epsilon \left(\langle D_x \left(D_I H(\tilde{x}_0(I;\mu), I; \mu) \right), \tilde{x}_1(I, \theta; \mu) \rangle \right. \\[2mm]
&\quad \left. + g^\theta(\tilde{x}_0(I;\mu), I, \theta; \mu, 0) \right) + \mathcal{O}(\epsilon^2),
\end{aligned}
\tag{7.31}
$$

where $\langle \cdot, \cdot \rangle$ represents the usual Euclidean inner product.

We want to study the dynamics of (7.31) near the resonance $I = I^r$. For this purpose we will change variables in order to derive a simpler equation that describes the dynamics in a neighborhood of the resonance. Substituting

$$
\begin{aligned}
I &= I^r + \sqrt{\epsilon} h, \tag{7.32} \\
\theta &= \theta \tag{7.33}
\end{aligned}
$$

into (7.31), Taylor expanding in powers of $\sqrt{\epsilon}$, and rescaling time by letting $\tau = \sqrt{\epsilon} t$ gives the equations

$$
\begin{aligned}
h' &= g^I + \sqrt{\epsilon} G(h, \theta, \mu) + \mathcal{O}(\epsilon), \\
\theta' &= \left(\langle D_x(D_I H), D_I \tilde{x}_0 \rangle + D_I^2 H \right) h + \sqrt{\epsilon} F(h, \theta, \mu) + \mathcal{O}(\epsilon),
\end{aligned}
\tag{7.34}
$$

where the prime denotes differentiation with respect to the rescaled time τ,

$$
G(h, \theta, \mu) = \left(\langle D_x g^I, D_I \tilde{x}_0 \rangle + D_I g^I \right) h
$$

and

$$
\begin{aligned}
F(h, \theta, \mu) &= \tfrac{1}{2}(\langle \langle (D_x(D_x D_I H)) D_I \tilde{x}_0, D_I \tilde{x}_0 \rangle + \langle D_x(D_I H), D_I^2 \tilde{x}_0 \rangle \\[2mm]
&\quad + 2\langle D_x(D_I^2 H), D_I \tilde{x}_0 \rangle + D_I^3 H) h^2 + \langle D_x(D_I H), \tilde{x}_1 \rangle + g^\theta,
\end{aligned}
\tag{7.35}
$$

and where all functions are evaluated at $\tilde{x}_0(I;\mu) = \tilde{x}_0(I^r;\mu)$, $I = I^r$, θ, μ, and $\epsilon = 0$. The important advantage gained in localizing the equations near the resonance is that at $\epsilon = 0$, (7.34) is a one-degree-of-freedom (hence, integrable) Hamiltonian system given by

$$h' = g^I = -D_\theta \mathcal{H},$$

$$\theta' = \left(\langle D_x(D_I H), D_I \tilde{x}_0 \rangle + D_I^2 H \right) h = D_h \mathcal{H},$$

(7.36)

where

$$\mathcal{H}(h, \theta) = \left(\langle D_x(D_I H), D_I \tilde{x}_0 \rangle + D_I^2 H \right) \frac{h^2}{2} - \int_{\theta_0}^{\theta} g^I d\bar{\theta}$$

is the Hamiltonian function. The integrable Hamiltonian structure at leading order is typical near resonances (see, e.g., Wiggins [1990]) and is extremely useful for understanding the qualitative (as well as the quantitative) structure of the dynamics near the resonance on $\mathcal{M}_\epsilon$. The $\sqrt{\epsilon}$ dependence in the change of variables given in (7.33) is a consequence of the implicit assumption that

$$\langle D_x(D_I H(\tilde{x}_0(I^r; \mu), I^r; \mu)), D_I \tilde{x}_0 \rangle + D_I^2 H(\tilde{x}_0(I^r; \mu), I^r; \mu) \neq 0.$$

If this is violated, then a scaling with a different fractional power of ϵ is required; see Wiggins [1990] for details.

Returning now to our discussion of the foliation of $W^s(\mathcal{M}_\epsilon)$ and $W^u(\mathcal{M}_\epsilon)$ near the resonance, we rewrite (7.34) *without rescaling time*:

$$\dot{h} = \sqrt{\epsilon} g^I + \epsilon G(h, \theta, \mu) + \mathcal{O}(\epsilon^{\frac{3}{2}}),$$

$$\dot{\theta} = \sqrt{\epsilon} \left(\langle D_x(D_I H), D_I \tilde{x}_0 \rangle + D_I^2 H \right) h + \epsilon F(h, \theta, \mu) + \mathcal{O}(\epsilon^{\frac{3}{2}}).$$

(7.37)

Note that for $\epsilon = 0$, (7.37) reduces to

$$\dot{h} = 0, \tag{7.38}$$

$$\dot{\theta} = 0. \tag{7.39}$$

Of course, the dynamics in the full phase space are described by

$$\dot{x} = J D_x H(x, I^r; \mu) + \sqrt{\epsilon} D_I(J D_x H(x, I^r; \mu)) h$$

$$+ \tfrac{\epsilon}{2} D_I^2 (J D_x H(x, I^r; \mu)) h^2 + \epsilon g^x(x, I^r, \theta; \mu, 0) + \mathcal{O}(\epsilon^{3/2}),$$

$$\dot{h} = \sqrt{\epsilon} g^I(x, I^r, \theta; \mu, 0) + \epsilon D_I g^I(x, I^r, \theta; \mu, 0) h + \mathcal{O}(\epsilon^{3/2}),$$

(7.40)

$$\dot{\theta} = D_I H(x, I^r; \mu) + \sqrt{\epsilon} D_I^2 H(x, I^r; \mu) h + \tfrac{\epsilon}{2} D_I^3 H(x, I^r; \mu) h^2$$

$$+ \epsilon g^\theta(x, I^r, \theta; \mu, 0) + \mathcal{O}(\epsilon^{3/2}).$$

For $\epsilon = 0$, (7.40) reduces to

$$\dot{x} = J D_x H(x, I^r; \mu),$$

$$\dot{h} = 0, \qquad\qquad (7.41)$$

$$\dot{\theta} = D_I H(x, I^r; \mu).$$

There are several features that we want to point out concerning the above sets of equations.

1. From (7.41), we see that at $\epsilon = 0$ the neighborhood of the resonance on $\mathcal{M}$ *in the variables scaled as in (7.33)* consists entirely of fixed points. We can think of the change of variables in (7.33) as "blowing up" the circle of fixed points into an annulus of fixed points centered at $I = I^r$.

   We make this notation more precise. Let

   $$\mathcal{A}_\epsilon = \left\{ (x, h, \theta) \mid x = \tilde{x}_\epsilon(I^r + \sqrt{\epsilon}h, \theta; \mu), \mid h \mid < C, \theta \in S^1 \right\}$$

   and

   $$\mathcal{A} = \left\{ (x, h, \theta) \mid x = \tilde{x}_0(I^r + \sqrt{\epsilon}h; \mu), \mid h \mid < C, \theta \in S^1 \right\},$$

   where the constant C is chosen large enough to contain the trajectories of interest. We refer to $\mathcal{A}$ as the *unperturbed annulus* and it should be clear that it consists entirely of fixed points for (7.41). The stable and unstable manifolds of $\mathcal{A}$, denoted $W^s(\mathcal{A})$ and $W^u(\mathcal{A})$, are obtained by restricting the I values in $W^s(\mathcal{M})$ and $W^u(\mathcal{M})$ to $I = I^r + \sqrt{\epsilon}h, \mid h \mid < C$.

   With the I value suitably restricted, the persistence theorem 7.5.1 can be applied to $\mathcal{A}$ and (7.41), with the conclusion that $\mathcal{A}$, along with its stable and unstable manifolds, persists as $\mathcal{A}_\epsilon$.

2. For ϵ small, but nonzero, we see that, roughly speaking, the character of the dynamics in the x variables is not altered much under the influence of the perturbation. Theorem 7.5.1 makes this more precise. However, the dynamics on the annulus is radically different. Indeed, for $\epsilon = 0$ there is no dynamics on the annulus (it consists entirely of fixed points), whereas for ϵ small, the typical resonance structure is created as was revealed through a study of (7.37) under the rescaled time or, *slow time*, $\tau = \sqrt{\epsilon}t$.

As we mentioned earlier, we want to relate the asymptotic behavior of trajectories in the stable and unstable manifolds of the annulus to trajectories in the annulus for ϵ small, but nonzero. From Eqs. (7.37)–(7.41) as well as the discussion following these equations it should be clear that this is a singular perturbation problem; however, since we will require infinite time results, classical singular perturbation approaches will be of limited use. Rather, the problem is most naturally addressed from the geometrical, dynamical systems viewpoint that utilizes the *foliation* of the stable and unstable manifolds by submanifolds consisting of initial conditions of trajectories that have the same "asymptotic phase." This is stated more precisely in the following foliation theorem that we formulate in the context of Eq. (7.40).

Theorem 7.5.2 *There exists $\delta_0 > 0$ and $\epsilon_0 > 0$ such that given any point $(\bar{h}, \bar{\theta}) \in \mathcal{A}_\epsilon$ there exists a family of one-dimensional curves, called the **stable fibers**, that can be represented as graphs as follows:*

$$x_2 = x_2(x_1; \bar{h}, \bar{\theta}, \mu, \sqrt{\epsilon}),$$

$$h = h(x_1; \bar{h}, \bar{\theta}, \mu, \sqrt{\epsilon}), \tag{7.42}$$

$$\theta = \theta(x_1; \bar{h}, \bar{\theta}, \mu, \sqrt{\epsilon}),$$

*where $x \equiv (x_1, x_2)$. The point $(\bar{h}, \bar{\theta})$ is referred to as the **basepoint** of the fiber. These graphs are defined for any $0 < \delta \leq \delta_0, 0 < \epsilon \leq \epsilon_0$ with $|x_1| \leq \delta$ and $\mu \in \mathbb{R}^p$. Moreover, these curves have the following properties:*

1. *They are C^r in x_1 and C^r in $(\bar{h}, \bar{\theta}, \mu, \sqrt{\epsilon})$.*

2. $x_2(\tilde{x}_{1\epsilon}(I^r + \sqrt{\epsilon}\bar{h}, \bar{\theta}, \mu); \bar{h}, \bar{\theta}; \mu, \sqrt{\epsilon}) = \tilde{x}_{2\epsilon}(I^r + \sqrt{\epsilon}\bar{h}, \bar{\theta}, \mu),$ (7.43)

   $h(\tilde{x}_{1\epsilon}(I^r + \sqrt{\epsilon}\bar{h}, \bar{\theta}, \mu); \bar{h}, \bar{\theta}, \mu, \sqrt{\epsilon}) = \bar{h},$

   $\theta(\tilde{x}_{1\epsilon}(I^r + \sqrt{\epsilon}\bar{h}, \bar{\theta}, \mu); \bar{h}, \bar{\theta}, \mu, \sqrt{\epsilon}) = \bar{\theta},$

   where, recall, $\mathcal{A}_\epsilon$, with the I values suitably restricted, is the graph of $\tilde{x}_\epsilon(I, \theta, \mu) \equiv (\tilde{x}_{1\epsilon}(I, \theta, \mu), \tilde{x}_{2\epsilon}(I, \theta, \mu))$ (cf. Theorem 7.5.1).

3. $W^s_{loc}(\mathcal{A}_\epsilon) = \bigcup_{(\bar{h}, \bar{\theta}) \in \mathcal{A}_\epsilon} \left[x_2(x_1; \bar{h}, \bar{\theta}, \mu, \sqrt{\epsilon}) \right].$

4. *Let $(\bar{h}(t), \bar{\theta}(t))$ be a trajectory in $\mathcal{A}_\epsilon$ satisfying $(\bar{h}(0), \bar{\theta}(0)) = (\bar{h}, \bar{\theta})$ and let $(x_1(t), x_2(t), h(t), \theta(t))$ be a trajectory in $W^s_{loc}(\mathcal{A}_\epsilon)$ satisfying*

$$x_2(0) = x_2(x_1(0); \bar{h}, \bar{\theta}, \mu, \sqrt{\epsilon}),$$

$$h(0) = h(x_1(0); \bar{h}, \bar{\theta}, \mu, \sqrt{\epsilon}), \tag{7.44}$$

$$\theta(0) = \theta(x_1(0); \bar{h}, \bar{\theta}, \mu, \sqrt{\epsilon}),$$

i.e., the trajectory starts on the fiber with basepoint $(\bar{h}, \bar{\theta})$; *then*

$$| (x(t), h(t), \theta(t)) - (\tilde{x}_\epsilon(I^r + \sqrt{\epsilon}\bar{h}(t), \bar{\theta}(t), \mu), \bar{h}(t), \bar{\theta}(t)) | < Ce^{-\lambda t},$$

for all $t > 0$ *and for some* C, $\lambda > 0$ *as long as* $(\bar{h}(t), \bar{\theta}(t)) \in \mathcal{A}_\epsilon$. *In other words, trajectories starting on a stable fiber asymptotically approach the trajectory in* $\mathcal{A}_\epsilon$ *that starts on the basepoint of the same fiber, as long as the trajectory through this basepoint remains in* $\mathcal{A}_\epsilon$.

5. *The family of fibers form an invariant family in the sense that fibers map to fibers under the time-T flow map. Analytically, this is expressed as follows. Suppose* $(x_1(t), x_2(t), h(t), \theta(t))$ *is a trajectory satisfying*

$$x_2(0) = x_2(x_1(0); \bar{h}, \bar{\theta}, \mu, \sqrt{\epsilon}),$$

$$h(0) = h(x_1(0); \bar{h}, \bar{\theta}, \mu, \sqrt{\epsilon}), \tag{7.45}$$

$$\theta(0) = \theta(x_1(0); \bar{h}, \bar{\theta}, \mu, \sqrt{\epsilon});$$

then

$$x_2(t) = x_2(x_1(t); \bar{h}(t), \bar{\theta}(t), \mu, \sqrt{\epsilon}),$$

$$h(t) = h(x_1(t); \bar{h}(t), \bar{\theta}(t), \mu, \sqrt{\epsilon}), \tag{7.46}$$

$$\theta(t) = \theta(x_1(t); \bar{h}(t), \bar{\theta}(t), \mu, \sqrt{\epsilon}).$$

6. *At* $\epsilon = 0$ *the unperturbed fibers correspond to the unperturbed heteroclinic orbits. Hence, the perturbed and unperturbed fibers are* $C^r \sqrt{\epsilon}$-*close.*

We make the following remarks concerning this theorem.

Remark 1. An identical result follows for the fibering of $W^u(\mathcal{A}_\epsilon)$.

Remark 2. Quasilinear partial differential equations whose solutions are the fibers can be derived. These equations are analogous to those given following Theorem 7.5.1. We will not require these for our calculations; however, the reader can find these equations in Fenichel [1979] where a general geometric singular perturbation theory is developed.

References

Abraham, R., Marsden, J. E. [1978] *Foundations of Mechanics.* Addison-Wesley.

Abraham, R. A., Marsden, J. E., and Ratiu, T. S. [1988] *Manifolds, Tensor Analysis, and Applications.* Second edition. Springer-Verlag: New York, Heidelberg, Berlin.

Allen, J. S., R. M. Samelson, and P. A. Newberger [1991] Chaos in a model of forced quasiperiodic flow over topography—an application of Melnikov's method. *J. Fluid Mech.,* **226**, 511–547.

Arnold, V. I. [1973] *Ordinary Differential Equations.* M.I.T. Press: Cambridge, MA.

Ball, J. [1973] Saddle point analysis for an ordinary differential equation in a Banach space, and an application to dynamic buckling of a beam, in *Nonlinear Elasticity* (R. W. Dickey, ed.), Academic Press: New York, 93–160.

Ball, J. M., Holmes, P. J., James, R. D., Pego, R. L, and Swart, P. J. [1991] On the Dynamics of Fine Structure. *J. Nonlinear Sci.,* **1**(1), 17–70.

Bates, P. W. and Jones, C. K. R. T. [1989] Invariant Manifolds for Semilinear Partial Differential Equations. *Dynamics Reported,* **2**, 1–38.

Benedicks, M. and Carleson, L. [1991] The Dynamics of the Hénon Map. *Annals of Math.,* **133**(1), 73–170.

Benettin, G., Galgani, L., Giorgilli, A., and Strelcyn, J.-M. [1984] A proof of Kolmogorov's theorem on invariant tori using canonical transformations defined by the Lie method. *Nuovo Cimento B,* **79**(2), 201–223.

Berger, M. and Gostiaux, B. [1988] *Differential Geometry: Manifolds, Curves, and Surfaces.* Springer-Verlag: New York, Heidelberg, Berlin.

Bleher, S., Grebogi, C., and Ott, E. [1990] Bifurcation to Chaotic Scattering. *Physica D,* **46**, 87–121.

Bruhn, B. and Koch, B. P. [1991] Heteroclinic bifurcations and invariant manifolds in rocking block dynamics. *Z. Natur-forsch.*, **46a**, 481–490.

Chow, S. N. and Hale, J. K. [1982] *Methods of Bifurcation Theory.* Springer-Verlag: New York, Heidelberg, Berlin.

Constantin, P., Foias, C., Nicolaenko, B., and Temam, R. [1989] *Integral Manifolds and Inertial Manifolds for Dissipative Partial Differential Equations.* Springer-Verlag: New York, Heidelberg, Berlin.

Coppel, W. A. [1978] *Dichotomies in Stability Theory.* Lecture Notes in Mathematics, Vol. 629. Springer-Verlag: New York, Heidelberg, Berlin.

Davis, M. J. [1987] Phase space dynamics of bimolecular reactions and the breakdown of transition state theory. *J. Chem. Phys.*, **86**(7), 3978–4003.

de la Llave, R. and Wayne, C. E. [1993] *On Irwin's proof of the pseudostable manifold theorem.* Univ. of Texas preprint.

Dieudonné, J. [1960] *Foundations of Modern Analysis.* Academic Press: New York.

Falzarano, J., Shaw, S. W., and Troesch, A. W. [1992] Applications of global methods for analyzing dynamical systems to ship rolling motion and capsizing. *International J. of Bifurcations and Chaos*, **21**, 101–115.

Fenichel, N. [1970] Ph. D. Thesis, New York University.

Fenichel, N. [1971] Persistence and smoothness of invariant manifolds for flows. *Ind. Univ. Math. J.*, **21**, 193–225.

Fenichel, N. [1974] Asymptotic stability with rate conditions. *Ind. Univ. Math. J.*, **23**, 1109–1137.

Fenichel, N. [1977] Asymptotic stability with rate conditions, II. *Ind. Univ. Math. J.*, **26**, 81–93.

Fenichel, N. [1979] Geometric singular perturbation theory for ordinary differential equations. *J. Diff. Eqns.*, **31**, 53–98.

Fuks, D. B. and Rokhlin, V. A. [1984] *Beginner's Course in Topology: Geometric Chapters.* Springer-Verlag: New York, Heidelberg, Berlin.

Gillilan, R. and Ezra, G. S. [1991] Transport and turnstiles in multidimensional Hamiltonian mappings for unimolecular fragmentation: application to van der Waals predissociation. *J. Chem. Phys.*, **94**(4), 2648–2668.

Guillemin, V. and Pollack, A. [1974] *Differential Topology.* Prentice Hall, Inc.: Englewood Cliffs, NJ.

Hadamard, J. [1901]. Sur l'iteration et les solutions asymptotiques des equations différentielles. *Bull. Soc. Math. France,* **29**, 224–228.

Hale, J. [1980] *Ordinary Differential Equations.* Robert E. Krieger Publishing Co., Inc.: Malabar, Florida.

Haller, G. and Wiggins, S. [1992] Whiskered tori and chaos in resonant Hamiltonian normal forms. Accepted for publication in the proceedings of the workshop on *Normal Forms and Homoclinic Chaos.* The Fields Institute: Waterloo, Ontario.

Haller, G. and Wiggins, S. [1993a] Orbits homoclinic to resonances: The Hamiltonian case. *Physica D*, **66**, 298–346.

Haller, G. and Wiggins, S. [1993b] N-Pulse homoclinic orbits in perturbations of hyperbolic manifolds of Hamiltonian equilibria. submitted to *Arch. Rat. Mech. Anal.*.

Henry, D. [1981]. *Geometric Theory of Semilinear Parabolic Equations.* Springer Lecture Notes in Mathematics Vol. 840. Springer-Verlag: New York, Heidelberg, Berlin.

Hirsch, M.W., Pugh, C.C., and Shub, M. [1977] *Invariant Manifolds*, Lecture Notes in Mathematics Vol. 583. Springer-Verlag: New York, Heidelberg, Berlin.

Hirsch, M. W. and Smale, S. [1974] *Differential Equations, Dynamical Systems, and Linear Algebra.* Academic Press: New York.

Hoveijn, I. [1992] *Aspects of Resonance in Dynamical Systems.* Ph.D. thesis, University of Utrecht.

Irwin, M. C. [1970] On the stable manifold theorem. *Bull. London Math. Soc.*, **2**, 196–198.

Irwin, M. C. [1980] A new proof of the pseudostable manifold theorem. *Jour. London Math. Soc.*, **21**, 557–566.

Kang, I. S. and Leal, L. G. [1990] Bubble dynamics in time-periodic straining flows. *J. Fluid Mech.*, **218**, 41–69.

Kirchgraber, U. and Palmer, K. J. [1990] *Geometry in the Neighborhood of Invariant Manifolds of Maps and Flows and Linearization.* Pitman Research Notes in Mathematics Series. Longman Scientific & Technical. Published in the United States with John Wiley & Sons, Inc.: New York.

Kopell, N. [1979] A geometric approach to boundary layer problems exhibiting resonance. *SIAM J. Appl. Math.*, **37**(2), 436–458.

Kopell, N. [1985] Invariant manifolds and the initialization problem for some atmospheric equations. *Physica D*, **14**, 203–215.

Kovačič, G. and Wiggins, S. [1992] Orbits homoclinic to resonances, with an application to chaos in a model of the forced and damped sine–Gordon equation, *Physica D*, **57**, 185–225.

Liapunov, A. M. [1947] *Probleme general de la stabilite du movement*. Princeton University Press: Princeton.

Lin, X.-B. [1989] Shadowing lemma and singularly perturbed boundary value problems. *SIAM J. Appl. Math.*, **49**(1), 26–54.

Llavona, J. G. [1986] *Approximation of Continuously Differentiable Functions*. North-Holland: Amsterdam.

Lochak, P. [1992] Canonical perturbation theory via simultaneous approximation. *Russian Math. Surveys*, **47**(6), 57–133.

Marsden, J. E. and Scheurle, J. [1987] The construction and smoothness of invariant manifolds by the deformation method. *SIAM J. Math. Anal.*, **18**(5), 1261–1274.

McLaughlin, D., Overman II, E.A., Wiggins, S., and Xiong, X. [1993] Homoclinic orbits in a four dimensional model of a perturbed NLS equation: A geometric singular perturbation study. submitted to *Dynamics Reported*.

Milnor, J. W. [1965] *Topology from the Differentiable Viewpoint*. University of Virginia Press: Charlottesville.

Nijmeijer, H. and van der Schaft, A. J. [1990] *Nonlinear dynamical control systems*. Springer-Verlag: New York, Heidelberg, Berlin.

Odyniec, M. and Chua, L. [1983] Josephson-junction circuit analysis via integral manifolds. *IEEE Trans. on Circ. and Systems*, **cas-30**(5), 308–320.

Odyniec, M. and Chua, L. [1985] Josephson-junction circuit analysis via integral manifolds. *IEEE Trans. on Circ. and Systems*, **cas-32**(1), 34–45.

Ottino, J.M. [1989] *The Kinematics of Mixing: Stretching, Chaos and Transport*. Cambridge University Press: Cambridge.

Palis, J. and Takens, F. [1993] *Hyperbolicity & Sensitive Chaotic Dynamics at Homoclinic Bifurcations*. Cambridge University Press: Cambridge.

Perron, O. [1928]. Über stabilität und asymptotisches verhalten der integrale von differentialgleichungssystem. *Math. Z.*, **29**, 129–160.

Perron, O. [1929] Über stabilität und asymptotisches verhalten der lösungen eines systems endlicher differenzengleichungen. *J. Reine Angew. Math.*, **161**, 41–64.

Perron, O. [1930] Die stabilitätsfrage bei differentialgleichungen. *Math. Z.*, **1930**, 703–728.

Petschel, G. and Geisel, T. [1991] Unusual manifold structure and anomalous diffusion in a Hamiltonian model for chaotic guiding center motion. *Physical Review A*, **44**(12), 7959–7967.

Pöschel, J. [1993] Nekhoroshev estimates for quasi-convex Hamiltonian systems. *Mathematische Zeitschrift*, **213**(2), 187–216.

Rudin, W. [1964] *Principles of Mathematical Analysis*. McGraw-Hill: New York.

Sacker, R. J. [1969] A perturbation theorem for invariant manifolds and Hölder continuity. *J. Math. Mech.*, **18**, 187–198.

Sacker, R. J. and Sell, G. R. [1974] Existence of dichotomies and invariant splittings for linear differential systems. *J. Diff. Eqns.*, **15**, 429–458.

Silnikov, L. P. [1967] On a Poincaré-Birkhoff problem. *Math. USSR Sb.*, **3**, 353–371.

Spivak, M. [1979] *Differential Geometry, Vol. I*. Second edition. Publish or Perish, Inc.: Wilmington, DE.

Temam, R. [1988] *Infinite Dimensional Dynamical Systems in Mechanics and Physics*. Springer-Verlag: New York, Heidelberg, Berlin.

Terman, D. [1992] The Transition form Bursting to Continuous Spiking in Excitable Membrane Models. *J. Nonlinear Sci.*,**2**(2), 135–182.

Touma, J. and Wisdom, J. [1993] The Chaotic Obliquity of Mars. *Science*, **259**, 1294–1297.

Wagenhuber, J., Geisel, T., Niebauer, P., and Obermair, G. [1992] Chaos and anomalous diffusion of ballistic electrons in lateral surface superlattices. *Physical Review B*, **45**(8), 4372–4383.

Wells, J. C. [1976] Invariant Manifolds of non-linear operators. *Pac. Jour. Math.*, **62**, 285–293.

Whitney, H. [1936] Differentiable Manifolds. *Ann. Math.*, **37**(3), 645–680.

Wiggins, S. [1988] *Global Bifurcations and Chaos – Analytical Methods*. Springer-Verlag: New York, Heidelberg, Berlin.

Wiggins, S. [1990] On the geometry of transport in phase space, I. Transport in k-degree-of-freedom Hamiltonian systems, $2 \leq k < \infty$, *Physica D*, **44**, 471–501.

Wiggins, S. [1992] *Chaotic Transport in Dynamical Systems*. Springer-Verlag: New York, Heidelberg, Berlin.

Yi, Y. [1993a] A generalized integral manifold thorem. *J. Diff. Eq.*, **102**(1), 153–187.

Yi, Y. [1993b] Stability of integral manifold and orbital attraction of quasiperiodic motion. *J. Diff. Eq.*, **103**(2), 278–322.

Index

Applied Mathematical Sciences

(continued from page ii)